BASIC RIFLE MARKSMANCHIP TRAINER'S GUIDE

United States Army Infantry School

Fredonia Books
Amsterdam, The Netherlands

Basic Rifle Marksmanship Trainer's Guide

by
US Army Infantry School

ISBN: 1-4101-0691-8

Copyright © 2004 by Fredonia Books

Reprinted from the 1986 edition

Fredonia Books
Amsterdam, The Netherlands
http://www.fredoniabooks.com

All rights reserved, including the right to reproduce
this book, or portions thereof, in any form.

CONTENTS

	Page
FOREWORD	1
INTRODUCTION	2
THE BASIC APPROACH	4
THE BASIC INGREDIENTS	6
BALLISTICS	10
THE RIFLE	24
BASIC SHOOTING FUNDAMENTALS	27
FIRING POSITIONS	32
ORGANIZATION FOR TRAINING	35
REMEDIAL TRAINING	35
ZEROING	35
SHOT GROUPING	39
SCALED SILHOUETTE TARGETS	43
DOWNRANGE FEEDBACK	47
FIELD FIRE	50
TARGET DETECTION	50
RECORD FIRE	50
APPENDIXES:	
APPENDIX A. Training Aids and Procedures	A-1
APPENDIX B. BRM POI Recapitulation	B-1
APPENDIX C. Bibliography	C-1

  Page

FIGURES

Figure 1. Serviceability firing checks. 7
Figure 2. Minute of angle. 11
Figure 3. Shot group width as a function of range. 13
Figure 4. Effects of gravity. 14
Figure 5. Bullet drop when rifle is boresighted. 15
Figure 6. Bullet rise and drop--M16 rifle zeroed for 250 meters. 16
Figure 7. Adjusted optimum aiming point to allow for the effects of gravity. 17
Figure 8. Trajectory--150, 250, and 300 meters (cradle firings). 19
Figure 9. Wind direction by the clock system. 20
Figure 10. Effect of a 10-mph wind on bullet's path. 21
Figure 11. Determining wind velocity by the flag method. 22
Figure 12. Determining wind velocity by the pointing method. 23
Figure 13. Adjusted aiming point--wind. 23
Figure 14. Targets used for test firing. 26
Figure 15. Sight alinement. 30
Figure 16. Rifle stress (handheld). 34
Figure 17. Trajectory--250-meter zero. 35
Figure 18. Trajectory--long-range sight. 36
Figure 19. 25-meter zeroing target for M16A1 rifle (with standard sights). 38
Figure 20. Shot groups form unsupported positions. 40
Figure 21. Shot groups from entry-level soldier firings. 42
Figure 22. 25-meter scaled silhouette targets (slow). 44
Figure 23. 25-meter scaled silhouette targets (timed fire). 46
Figure 24. 75-meter known distance target. 48
Figure 25. 175-meter known distance target. 49

BASIC RIFLE MARKSMANSHIP TRAINER'S GUIDE

FOREWORD

PURPOSE

The purpose of this Guide is to assist trainers involved with the teaching of basic rifle markmanship (BRM) in acquiring the knowledge and skills necessary to be a more effective instructor of basic shooting skills. Additionally, this Guide serves as an introduction to the revised Basic Rifle Marksmanship Program.

SCOPE

This Guide has been designed specifically for the instructor of entry-level marksmanship. All aspects of the newly developed BRM program are discussed. However, this Guide specifically addresses basic shooting fundamentals that apply to all small arms and is, therefore, a valuable source of information for all units with M16A1-armed personnel.

There is no intent that this Guide serve as a one-stop reference for all marksmanship training. Several areas, such as mechanical training, have not been included in recent research efforts as the material contained in current literature is considered adequate.

It will be noted that much information in this Guide is new. The rationale for each change is explained in detail. The fundamentals contained here will be included in change 3 of FM 23-9, M16A1 Rifle and Rifle Marksmanship.

HOW TO USE THIS GUIDE

Any individual involved with marksmanship instruction should initially read the entire Guide and then review appropriate portions of the Guide before specific training periods. The Guide is organized to facilitate this type of use.

PREPARATION OF GUIDE

This document was prepared under Contract Number MDA 90379-M-7322 by Mellonics Systems Development Division of Litton Systems, Inc.; the US Army Research Institute, Fort Benning Field Unit; and the US Army Infantry School.

RECOMMENDED CHANGES

This Guide represents several changes from previous marksmanship instruction. It is also the first attempt at providing a publication for the trainer of basic marksmanship. The experience gained through field implementation will be invaluable in making this a better Guide. Recommended changes concerning the content of the instruction, the organization of this Guide, or other marksmanship ideas will be welcomed. All recommended changes will be fully evaluated by subject matter experts. The preaddressed tear-out page at the back of this book can be used, or address recommendations to:

```
                    Commandant
                    United States Army Infantry School
                    ATTN:  ATSH-I-V-SD
                    Fort Benning, GA  31905
```

INTRODUCTION

HISTORY OF AMERICAN MARKMANSHIP

During our early days, the US frontiersman became world renowned with his rifle. He could accomplish amazing feats compared to other marksmen. Although the frontiersman learned his skills from childhood and kept them current hunting game and fowl, marksmanship was considered an inborn trait. Consequently, sharpshooters for the Army were selected, rather than trained.

The theory of "natural marksmanship" was phased out with the passing of the frontier. The Army established the School of Musketry, where systematic training was given to marksmanship instructors. The school of Musketry, and later the Infantry School, produced well-qualified instructors who were able to mass-train infantrymen in rifle skills for World War I. During World War II, it was found that the hastily trained riflemen had difficulty with obscure targets and often did not fire at all. Aimed fire became less common and many engagements were decided by sheer weight of ammunition. Again, in Korea, combat troops developed the tendency to substitute quantity for quality, choosing to fire a large volume of ammunition instead of aimed fire. The Korean War brought renewed emphasis on marksmanship. In Vietnam, with the introduction of the fully automatic M16 rifle, the use of well-aimed fire was almost nonexistent, and the ability of the American soldier to hit the enemy at combat ranges continued to deteriorate.

THE COURSE OF FIRE

Two basic approaches have been employed by the US Army to teach basic marksmanship: the known distance (KD) approach and the TRAINFIRE approach. The KD approach (still employed by the US Marine Corps), used by the Army until the late 1950's, was characterized by firing at bull's-eye targets located at selected known distances (200, 300, and 500 yards) on an open range. The location of each bullet strike was marked with spotters by individuals in the "pits."

Through analysis of combat actions in Korea, it was found that many soldiers had difficulty applying the firing skills they had learned on the KD range to the engagement of combat targets--there were no bull's-eye targets on the battlefield.

With the recognition of the importance of the transfer of skills learned in training to those used in combat, it was considered necessary to examine the realism of the current training in this respect. TRAINFIRE was an attempt to develop and evaluate a rifle marksmanship training program designed for maximum rapid transfer to combat conditions. The original TRAINFIRE program as implemented had 80 hours of instruction, with many hours of field firing at pop-up targets. It also contained many trips back to the 25-meter line where fundamentals were reinforced and exact bullet strike locations could be examined. Since the adoption of the TRAINFIRE concept, as a result of 1954 studies, the field fire orientation of basic rifle marksmanship has changed little. However, major changes have occured in the amount of time allowed for training, the number of rounds of ammunition allocated for live firing, and the quality of downrange feedback (where bullets hit). In general, the trend has been to allocate either fewer hours or rounds, or both, to each phase of training. The time allocated to the fundamentals of shooting at 25 meters was reduced considerably and, except for zeroing, all knowledge of precise bullet strike was lost from the program.

Even a casual observation of soldiers undergoing the most recent TRAINFIRE program revealed major problems. The soldier was overloaded with knowledge requirements, was given inadequate time for skill practice, was given only one opportunity to master a complex zeroing procedure, and was expected to develop shooting skills without knowledge of performance. Therefore, a major thrust of the new program is to simplify procedures, focus only on the critical skills required to hit targets, provide adequate feedback of shooting performance, and eliminate confusion. In effect, the best aspects of the KD program and the TRAINFIRE program have been combined for presentation within current constraints of time and money.

DETERIORATION OF SHOOTING SKILLS

The American soldier does not shoot as well as he is capable. This statement is generally accepted throughout the United States Army. The shooting skills of the American soldier have deteriorated over the years and are considered to be at an unacceptably low level today. No conscious decisions have been made which directly contribute to this deterioration of shooting skills; however, many factors have had influence on this situation.

Many high priority and important subjects have been in competition for the time previously devoted to marksmanship training. The introduction of mass destruction weapons, antitank guided missiles, and other complex and highly sophisticated weapons and equipment tend to detract from the importance once placed on rifle marksmanship.

The continuing requirement to conduct training in the minimum amount of time at minimum cost has also had influence on the marksmanship program.

The transfer from KD firing to TRAINFIRE and subsequent modifications to TRAINFIRE, which resulted in less knowledge of downrange performance, added to the deterioration of marksmanship skills without it being obvious to the average training observer. The lack of downrange feedback and the lack of knowledge concerning bullet strike made it easy to ignore the problem. Standards were adjusted downward to allow for bad shooting and soldiers have been given the "benefit of the doubt" to insure that a high qualification rate was attained.

Several events have occurred which have taken the rifle out of the soldier's hand: physical training exercises with the rifle are no longer conducted, bayonet training and associated assault courses using the rifle are seldom conducted, dismounted drill with the rifle has been reduced, and security requirements have been increased—resulting in soldiers being able to handle the rifle only during official training time. The soldier is no longer familiar with his basic weapon.

Many factors have thus contributed to the deterioration of shooting skills. However, currently a major effort is underway to improve the marksmanship skills of the American soldier. This Guide and the revised Basic Rifle Marksmanship Program constitute tangible evidence of this effort. More importantly, this effort has the support of commanders at the highest echelons. The success of this effort hinges on the contributions of you—the marksmanship trainer. Your dedication to this effort will insure that the American soldier will again be known for his expertise with the service rifle.

IMPORTANCE OF BASIC RIFLE MARKSMANSHIP

"Many current Army regulations and policies place insufficient emphasis on individual, crew, and unit marksmanship. If the fighting Army does nothing else, we must be able to hit our targets. Conversely, if we do all other things right, but fail to hit and kill targets, we shall lose." This is the lead sentence in the US Army Vice Chief of Staff's 11 Dec 1980 Marksmanship memorandum. Over the years, the markmanship ability of an individual has been a predictor of his overall worth as a soldier in combat. Additionally, the collective marksmanship ability of its soldiers has been an accurate predictor of the overall worth of an army.

Soldiers, regardless of their job, must ultimately rely on the rifle for individual protection—perhaps to save their own lives. All soldiers will function more effectively when they are confident that their skill with the service rifle will insure their personal protection.

The combat power that can be generated by American Armed Forces is staggering. Unfortunately, the full force of this combat power could not to be applied in Vietnam nor can it be applied in other limited conflicts that may arise. To insure the security of the nation, we must be prepared for all-out war; however, the actual involvement of combat forces may be more closely related to a "rifle versus rifle" environment than a high intensity all-out war. It is difficult to visualize any military conflict that will not require all soldiers to be proficient with the service rifle, given the high mobility of all forces, the fluid environment expected on the battlefield, and the inability to provide completed security in rear areas. Therefore, the requirement for American soldiers to be proficient with the service rifle takes on added importance.

THE MARKSMANSHIP TRAINER

The marksmanship trainer is an individual assigned as a committee group instructor, drill sergeant, squad leader, section leader, or platoon leader. In short, EVERY SUPERVISOR IN THE ARMY SHOULD BE A QUALIFIED RIFLE MARKSMANSHIP TRAINER. The term "trainer" is used in this Guide to indicate any individual that is involved with the teaching of marksmanship skills. The importance of your job as a marksmanship trainer cannot be overemphasized. The degree of marksmanship proficiency attained by the soldier is largely dependent on the teaching of marksmanship fundamentals during entry-level training, but proper refresher training and skill retention practice must be provided during all subsequent assignments to insure combat readiness.

THE BASIC APPROACH

It should be remembered that this Guide is directed at teaching <u>basic</u> marksmanship skills to every soldier in the Army. Therefore, the approach to basic marksmanship is designed to place emphasis on those few key factors that have major influence on bullet strike, while minimizing or not addressing those factors that influence bullet strike an inch or two. This approach also places emphasis on a high first round hit probability for all close-in targets. There are four primary reasons for this.

1. Threat analysis places most personnel targets within 200 meters, with the majority of targets being at 150 meters or closer.

2. A trained firer using an M16 rifle of average quality with standard service ammunition is capable of hitting <u>every</u> personnel target out to ranges of 200 meters.

3. Enemy personnel from 0 to 150 meters pose an immediate threat to the soldier's life. It is important for all soldiers to have full confidence in their ability to kill every enemy soldier they can see at this range.

4. Assistance in engaging personnel targets at ranges beyond 200 meters could be expected to come from indirect fire, air support, or other weapon systems.

The previous program consisted of a quick zero procedure and the engagement of targets out to 300 meters (most of which were missed) with no feedback on misses and no feedback on the location of hits. The current program is desigend to develop and improve skills at a slower pace, providing precise feedback on the first 75 rounds fired, and produce a soldier that is confident of hitting <u>all</u> close-in and most distant targets as well.

In general, the key features of the BRM program are:

1. Simplified marksmanship fundamentals which focus only on the factors that contribute most to hitting targets.

2. Simplified zeroing targets that result in fewer zeroing errors and promote an understanding of the zeroing process.

3. Firing at scaled silhouette targets on the 25-meter range, that provides for skill practice in engaging silhouette targets, with accurate feedback on hits and misses. This exercise requires the use of the long-range sight, which causes bullet impact to coincide with point of aim.

4. Downrange feedback (75 and 175 meters) that allows a check of the zero at distant targets; provides knowledge of bullet trajectory, effects of wind, effects of gravity; and includes effective practice in hold-off. When available, a standard KD range is used for this period.

5. Zero confirmation gives the soldier and trainer a final check of fundamental ability and instills confidence in the zero before the record fire portion of BRM. This exercise also permits the soldier to make any necessary minor adjustments.

6. Combat firing, conducted the morning of record fire, effectively improves soldier performance. This exercise tasks the soldier to use all the skills trained during BRM against a difficult engagement scenario.

7. A revised qualification course that increases the minimum hits required to qualify from 43 percent (17 of 40) to 58 percent (23 of 40).

To better understand the approach to basic marksmanship, assume that there are at least two separate and different set of skills associated with marksmanship--competitive marksmanship and combat shooting. Precision is required in the training of the competitive shooter. Armed with a rifle and ammunition capable of consistently shooting half minute of angle*, the competitive shooter is provided with unlimited time, ammunition, and facilities for the purpose of refining marksmanship skills. All skill practices are performed with immediate feedback of where each bullet hits. Any factor that causes the bullets to stray from a four-square-inch area on a distant target must be of concern to the competitive shooter. Accordingly, a list of fundamentals for the competitive shooter must include all factors that have the slightest influence on the strike of the bullet.

---

*Minute of angle is explained in the ballistics section.

A more general and less detailed program is required for the basic soldier learning combat marksmanship. The soldier is provided with a weapon and ammunition combination that may be capable of firing only three minutes of angle, and is provided with limited time, ammunition, and facilities to learn shooting skills. However, the soldier is not required to hit a four-square-inch area. The soldier receives maximum credit during record fire for hitting anywhere on a 700-square-inch target (E-type silhouette). Given the many constraints associated with teaching soldiers to shoot--lack of shooting experience of entry-level soldiers, limited time, ammunition, and facilities, and inadequate numbers of trained instructor personnel--it appears logical that the list of shooting fundamentals developed for the initial-entry soldier should be somewhat different than a list prepared for the competitive shooter.

The fundamentals previously taught to entry-level soldiers contained many factors that would be more appropriately introduced in advanced marksmanship or a competitive program. The average soldier has been overwhelmed with emphasis on several factors that have only minor influence on the strike of the bullet. Meanwhile, soliders have been missing half of the 700-square-inch targets. The basic approach is to conduct training which will give all soldiers confidence that they can hit combat targets.

THE BASIC INGREDIENTS

Any shooting performance represents the collective capability of the rifle, ammunition, and shooter. Before the contribution required of a soldier can be determined, the capabilities of the rifle and ammunition used by the soldier must be known. Because this information was not readily available in reliable form for typical rifles, firing tests[1] were conducted to determine the performance capability of the typical rifle. The requirements, training procedures, and standards in the current program are based on the performance characteristics of the typical M16A1 rifle firing typical service ammunition.

---

[1] See bibliography

AMMUNITION

The basic marksmanship trainer has little control over the quality of training ammunition. Ammunition purchased by the Army must be capable of firing a four-inch or smaller group at 200 yards; therefore, the ammunition is relatively accurate and should not be of concern to the basic marksmanship trainer. Some small difference in performance may be noted between lots; therefore, it would be ideal if all firing from zero to record fire was accomplished with the same lot of ammunition. Service rifle ammunition is manufactured in lots of one to three million rounds, so it is normal to find the same lot being used for a complete cycle of training.

RIFLE QUALITY

Unlike the ammunition, there are many variables associated with the rifle over which the marksmanship trainer has some control.

The local direct support maintenance facility can perform five basic serviceability checks. Any weapon not meeting specifications will be repaired or eliminated from the active inventory. Unfortunately, passing all serviceability checks is no guarantee that a rifle will shoot accurately. Therefore, any time a weapon that has passed direct support maintenance serviceability checks is suspected to be the cause of large shot groups, it should be fired by an experienced marksman. A record of shot group sizes, fired by an experienced marksman, has proven to be an indicator of future weapons performance. <u>It is, therefore, recommended that a record be maintained for each weapon assigned to a company.</u> This record should contain the average shot group size each weapon is capable of firing as well as the qualification scores fired by each weapon during previous firings. This procedure will eliminate the possibility that a bad shooting performance is caused by the weapon or ammunition, and will allow attention to be rightfully focused on the individual's application of shooting fundamentals.

The importance of devoting time to rifle quality is illustrated in figure 1. The top targets were fired with an M16 that passed all available serviceability checks (checks conducted on two separate occasions by the Army Marksmanship Unit and the Direct/General Support Small Arms Maintenance Shop at Fort Benning, Georgia), and the bottom targets were fired by the same weapon after the upper receiver was replaced. To insure this type of repair is made, however, someone must identify the weapon as a bad shooting weapon.

These four targets were fired by the same firer with the same rifle from the same supported position on a 25-meter range. Before firing the first two five-round groups (targets 1 and 2), the rifle passed basic serviceability checks: barrel straightness, headspace, bore erosion, muzzle erosion, and trigger pull. Additionally, a physical measurement of the bore, a check of torque readings, and the recording of muzzle velocity for several rounds found those measures to be well within specifications. The next two five-round groups (targets 3 and 4) were fired after the upper receiver was replaced. With knowledge that the rifle would not shoot accurately, the decision to replace the upper receiver was a judgement call. Remember, the only way to determine if this type of repair is needed is to identify the rifle as a bad shooting rifle. Note also that targets 1 and 2 have bullet holes which are not round. This indicates that the bullets had not yet stabilized when they hit the target, an indication of a bad weapon.

Figure 1. Serviceability firing checks.

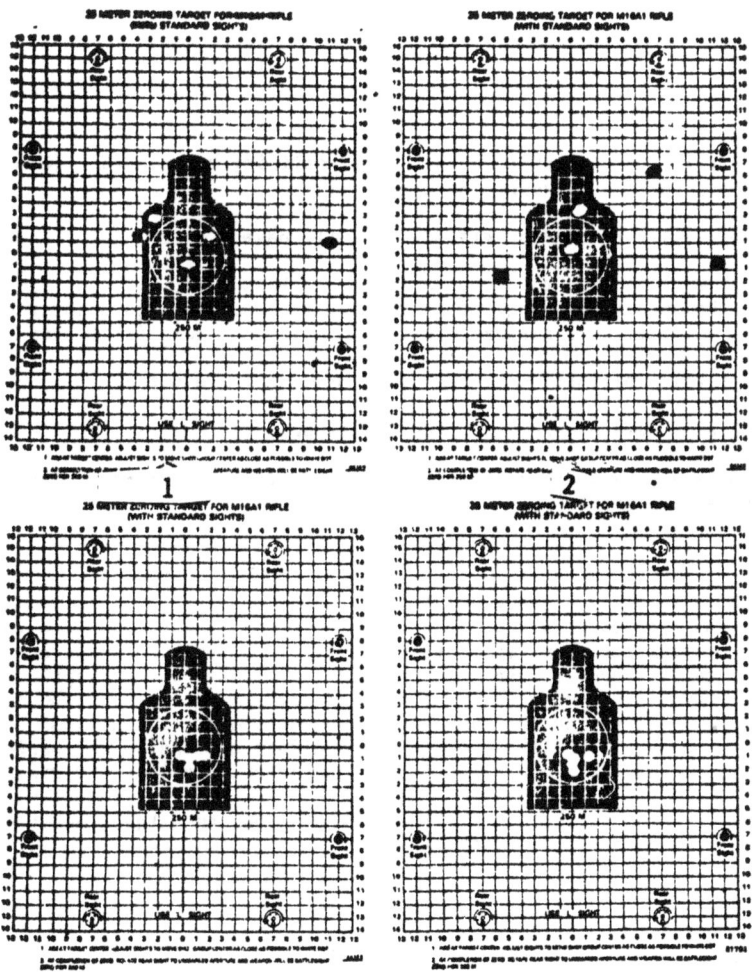

THE SOLDIER

With confidence that the rifles in the arms room are all of high quality and that good ammunition will be delivered to the firing range, attention can be turned to the soldier--who contributes the largest number of variables to the firing process. Good shooters are made, not born. It helps to have good eyes and good muscle coordination, but practically every soldier allowed to enlist in the Army is capable of becoming an expert with the M16A1 rifle. You may expect the following with a new group of entry-level soldiers$_2$:

---

^{2}See bibliography.

---

   1. Eighty-seven percent are right handed and will feel comfortable firing from the right shoulder.

   2. Eleven percent are left handed and will feel comfortable firing from the left shoulder.

   3. Two percent will not be sure of which shoulder they prefer to fire from. (These soldiers will not have a strong dominant hand, but is normally better to have them fire with the hand that matches their dominant eye.)

   4. A percent will have a dominant eye (app A) that is different from their dominant hand. (Individuals with a strong dominant hand will normally prefer to fire with their dominant hand and initially use a patch over their nonfiring eye.)

   5. Five percent will have difficulty keeping the nonfiring eye closed. (Some of these will learn faster with an eye patch on the nonfiring eye.)

   6. Twenty percent will need to wear glasses and will have a good pair of glasses.

   7. Twenty percent will be able to shoot better with glasses but will not have glasses or will not wear them during marksmanship training. (If you want your soldiers to shoot their best, this item should be checked. If they have glasses, insist that they be worn for training.)

   8. Twenty percent will initially have some fear of firing the rifle.

   9. Fifty percent will have no previous shooting experience.

   It is to your advantage to know your soldiers' traits before the start of formal marksmanship training and resolve potential problems.

BALLISTICS

The introduction of downrange feedback and other efforts to improve marksmanship require some knowledge of bullet flight characteristics. When the 5.56 projectile is launched into the earth's atmosphere at some 2,200 miles per hour (mph), it is acted on by various forces and elements; for example, gravity, wind, temperature, atmospheric pressure, humidity, and variability among bullets. There is enough variability associated with the firing of each round that you should not expect to see many shot groups with three bullets in the same hole. However, most of the items have negligible influence on the high velocity 5.56 bullet and are of no direct concern to the basic marksmanship trainer. Variability caused by the shooter is the most important variable and will be covered later.

There are three factors that influence bullet strike to such an extent that they must be discussed in detail: minute of angle, gravity, and wind.

MINUTE OF ANGLE

A minute of angle is the standard unit of measure used in adjusting rifle sights and many other ballistic related measurements or adjustments. It was used in a previous section of this Guide to indicate the accuracy of a rifle. Minute of angle is explained here to assist in understanding three things:

   1. How sight changes will effect bullets at range.

   2. How shot group size will increase at range.

   3. How a small movement of the rifle during firing will cause bullets to miss the target.

You may recall from map reading instruction that a circle is divided into 360 degrees. Each degree is further divided into 60 minutes, so that a circle containes 21,600 minutes (360 x 60). As indicated by figure 2, at a distance of 100 yards one minute of angle is equal to 1 inch. The rule to remember is that one minute of angle is equal to <u>1 inch at 100 yards</u>. As you can see in figure 2, when the circle is increased to a radius of 200 yards, the angle covers twice the distance it did at 100 yards, or 2 inches. And, of course, the rule applies as range increases--3 inches at 300 yards, 4 inches at 400 yards. Remember these three things:

1. One minute of angle is equal to 1 inch at 100 yards.

2. Once the angle is established, the width covered will increase in direct proportion to range--if it is 1 inch at 100 yards, it will be 2 inches at 200 yards, etc.

3. One click of adjustment on the M16 rifle, for windage or elevation, is equal to one minute of angle.

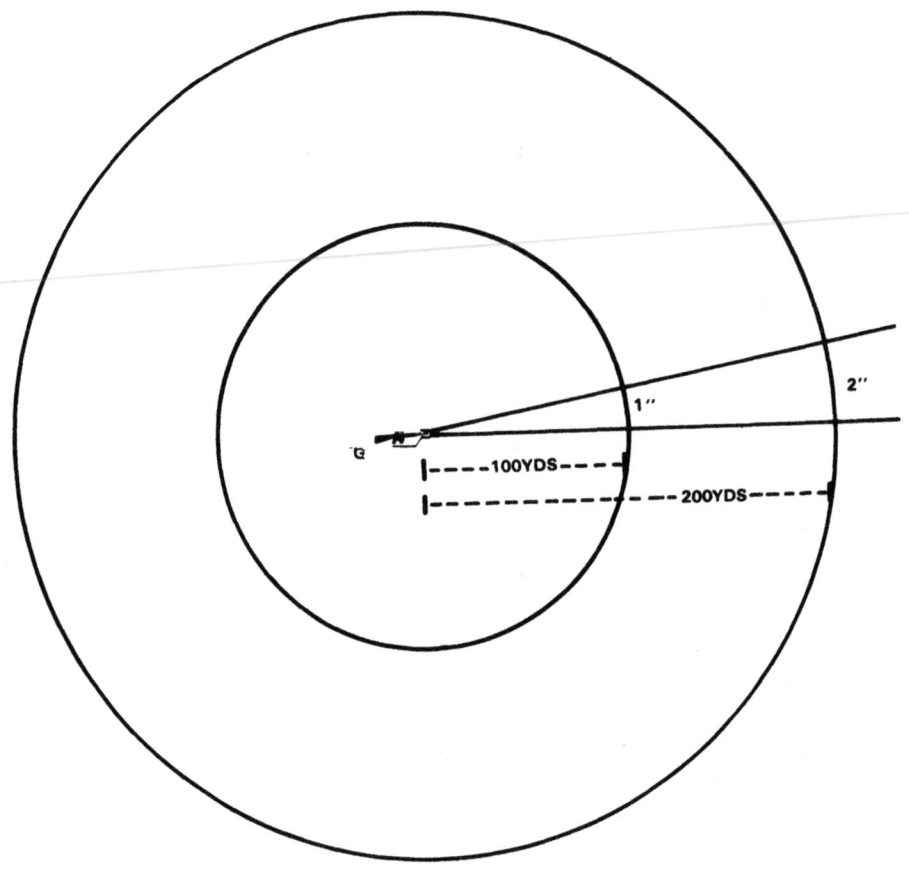

Each click of windage (rear sight) or elevation (front sight) adjustment on the M16 rifle is equal to one minute of angle, or 1 inch at 100 yards. Since the M16 sight radius (distance from the front to rear sight) is 19 3/4 inches, an actual sight movement of only 5/1000 inch moves the bullet strike 1 inch at 100 yards, 2 inches at 200 yards, etc. You can see that a small movement of the rifle during firing would cause a large error at distant targets.

Figure 2. Minute of angle.

When target ranges were measured in yards, it was simple to remember sight change informtion--1 inch at 100 yards, 1/4 inch at 25 yards, etc. Using the metric system, it makes it more difficult. However, since a meter is only 3 1/3 inches longer than a yard, the rule provides an idea of what effect sight changes will have at various ranges.

The metric system of measurement is now used for most measurements. This Guide uses inches several times because the US customary unit of measurement is more familiar. You should know the following:

1 inch	=	2 1/2 centimeters (2.54)
1 yard	=	.9 meters (.9144)
1 centimeter	=	.4 inches (.3938)
1 meter	=	1.1 yards (1.0937)
1 meter	=	100 centimeters
1 centimeter	=	10 millimeters

How far does one click of windage or elevation move the strike of the bullet at 25 meters? At 25 yards, it would be 1/4 inch, so at 25 meters (27.5 yards), it would be a little more than 1/4 inch, or .7 centimeters. Also note that .7 centimeters is the same as 7 millimeters.

Figure 3 depicts two bullets that are 1 inch apart at 25 meters. The same rule applies as to the increase in distance between bullets as range increases--if bullets are 2 inches apart at 50 meters, they will be 4 inches apart at 100 meters, 8 inches apart at 200 meters, etc.

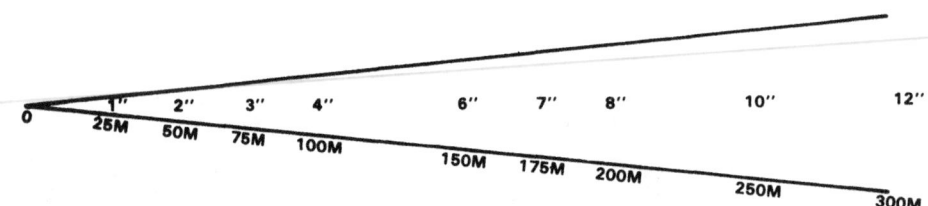

In effect, the angle created by these two bullets between the rifle and the 25-meter line will not change as the bullets continue downrange. Therefore, the bullets will be 2 inches apart at 50 meters, 3 inches apart at 75 meters, etc. At 175 meters, they will be 7 inches apart. Note that the shot groups at 175 meters will be seven times larger than they were at 25 meters. Just remember that the angles do not change; therefore, the range (175 meters) divided by the zero distance (25 meters) is equal to the expected increase in shot group size (7X).

Figure 3. Shot group width as a function of range.

GRAVITY

Since the M16 fires a high velocity round, it has what is often referred to as a "flat trajectory" from 0 to 300 meters. What this means is that when the M16 rifle is zeroed for 250 meters, the path of the bullet above or below the line of sight is less than that for most other military rifles. This is important because it allows targets from 0 to 300 meters to be hit using approximately the same point of aim. However, because of gravity, no bullet travels in a straight line.

Take any object that is similar in weight to a bullet (coin, key, etc.). Hold the object 24 inches above the floor and drop it—that is almost the same effect gravity has on a 5.56 bullet that is fired at a 300-meter target. The bullet is supported by the rifle until firing, but the moment the bullet leaves the muzzle, it is acted on by gravity in the same manner as a bullet you drop from you hand. (See figure 4.)

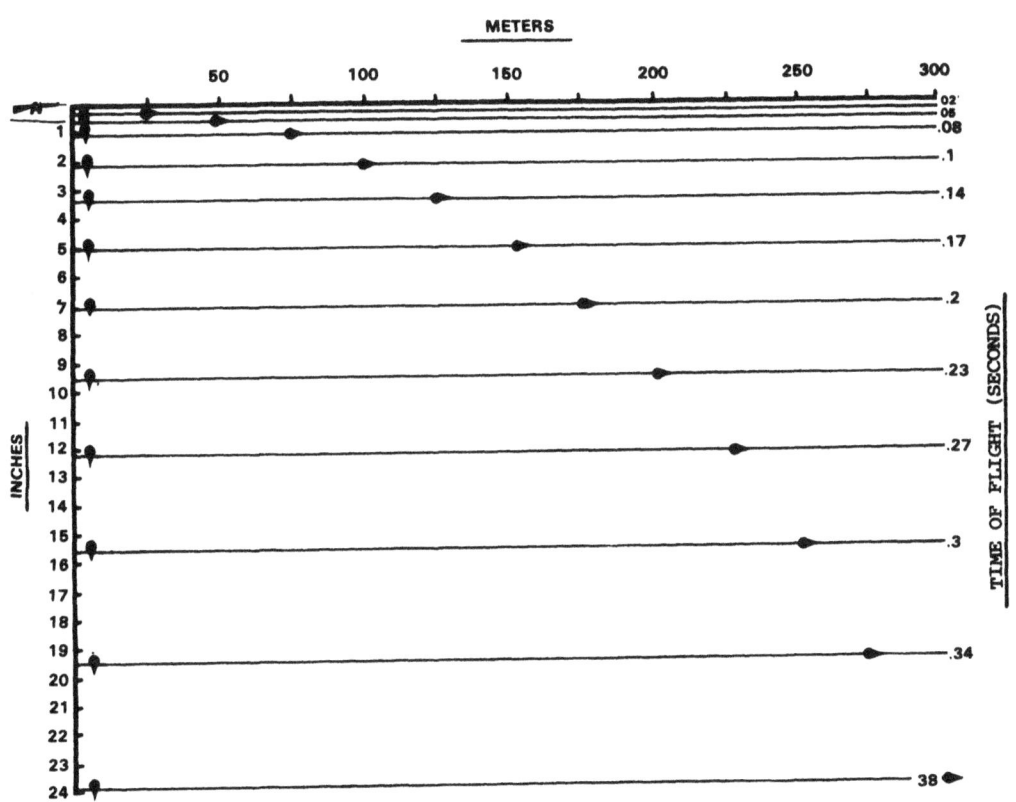

This chart shows that the effect of gravity on a bullet fired from a rifle is the same as if the bullet was dropped by hand. In a little more than 1/3 second the bullet falls 24 inches and travels 300 meters. The time bands are keyed to the fired bullet at 25-meter intervals--the lower bands are wider apart because the bullet loses velocity more rapidly as range increases.

Figure 4. Effects of gravity.

Figure 5 shows the bullet path for a rifle boresighted (the bore of the rifle pointed at the target, line of sight being through the bore) on a 300-meter target. As shown in figure 5, the bullet drops 16 inches by the time it reaches 250-meters or 24 inches at a distance of 300 meters.

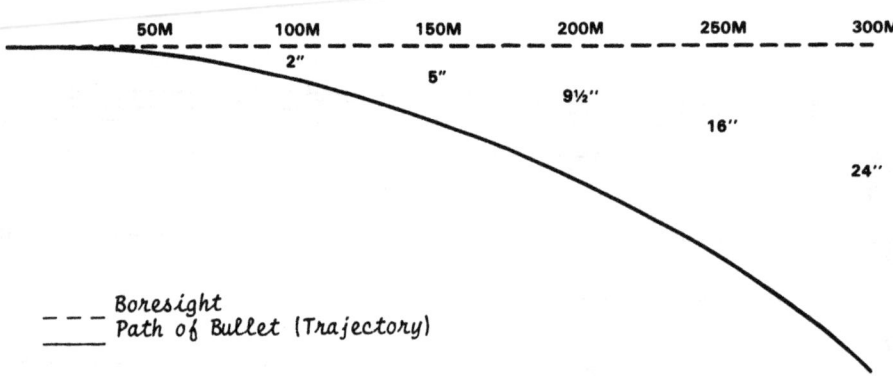

When the bore of the rifle is alined with a distant aiming point, the effects of gravity are significant--causing the bullet to drop 24 inches from boresight at 300 meters.

Figure 5. Bullet drop when rifle is boresighted.

It is important to note that gravity and wind (to be discussed in the following section) have an increasingly greater effect on bullets as range increases. As you can see, by dropping an object from a height of 24 inches, the object is traveling much faster the last inch of the distance than it is the first inch. Wind works in a similar way. The 5.56 bullet has lost over 36 percent of its initial speed (muzzle velocity) by the time it reaches 300 meters. Since gravity and wind have an increasing effect on the bullet, they will naturally have even more of an effect at greater ranges because it takes the bullet longer to travel a given distance.

It is obvious that something must be done to overcome the effects of gravity and make the engagement of targets out to 300 meters as simple as possible. A study of M16 trajectory data reveals that setting the sights to hit at 250 meters is the best compromise for hitting all targets from 25 to 300 meters. Figure 6 shows the bullet close to the line of sight (looking through the rifle sights) at all ranges.

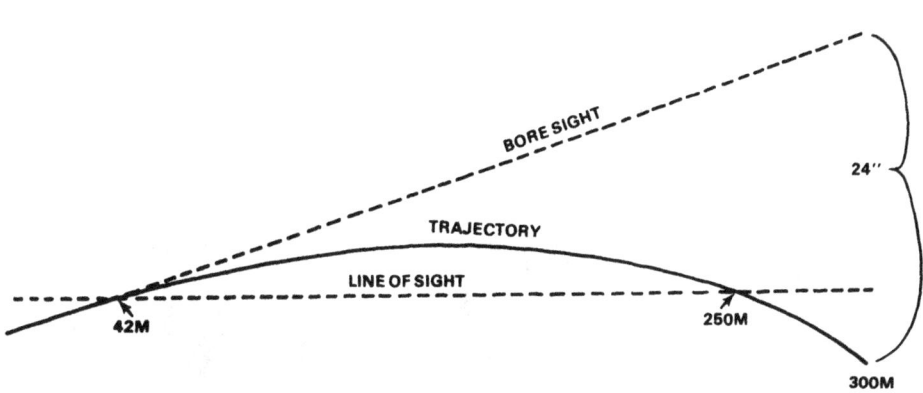

The bore of the M16 is 2.6 inches below the front sight, therefore, the bullet starts 2.6 inches below the line of sight. It crosses line of sight at a distance of 42 meters. It reaches its maximum height above line of sight approximately two-thirds the distance to the target (250 meters). Exact measurements are shown in figure 17.

Figure 6. Bullet rise and drop--M16 rifle zeroed for 250 meters.

With this sight setting, targets between 50 and 300 meters can be hit by aiming at the center of all targets. Point of aim will be discussed in detail later. The placement of the front sight that will provide the highest probability of getting all rounds on the target is illustrated in figure 7. New firers may perform better by aiming for the center of all targets.

The placement of the front sight as shown will provide the highest probability of hitting all targets with all bullets when there is no wind.

Figure 7. Adjusted optimum aiming point to allow for the effects of gravity.

The targets in figure 8 show the results of a center-of-mass aiming point for ranges of 150, 250 and 300 meters for rifles battlesight zeroed for 250 meters.

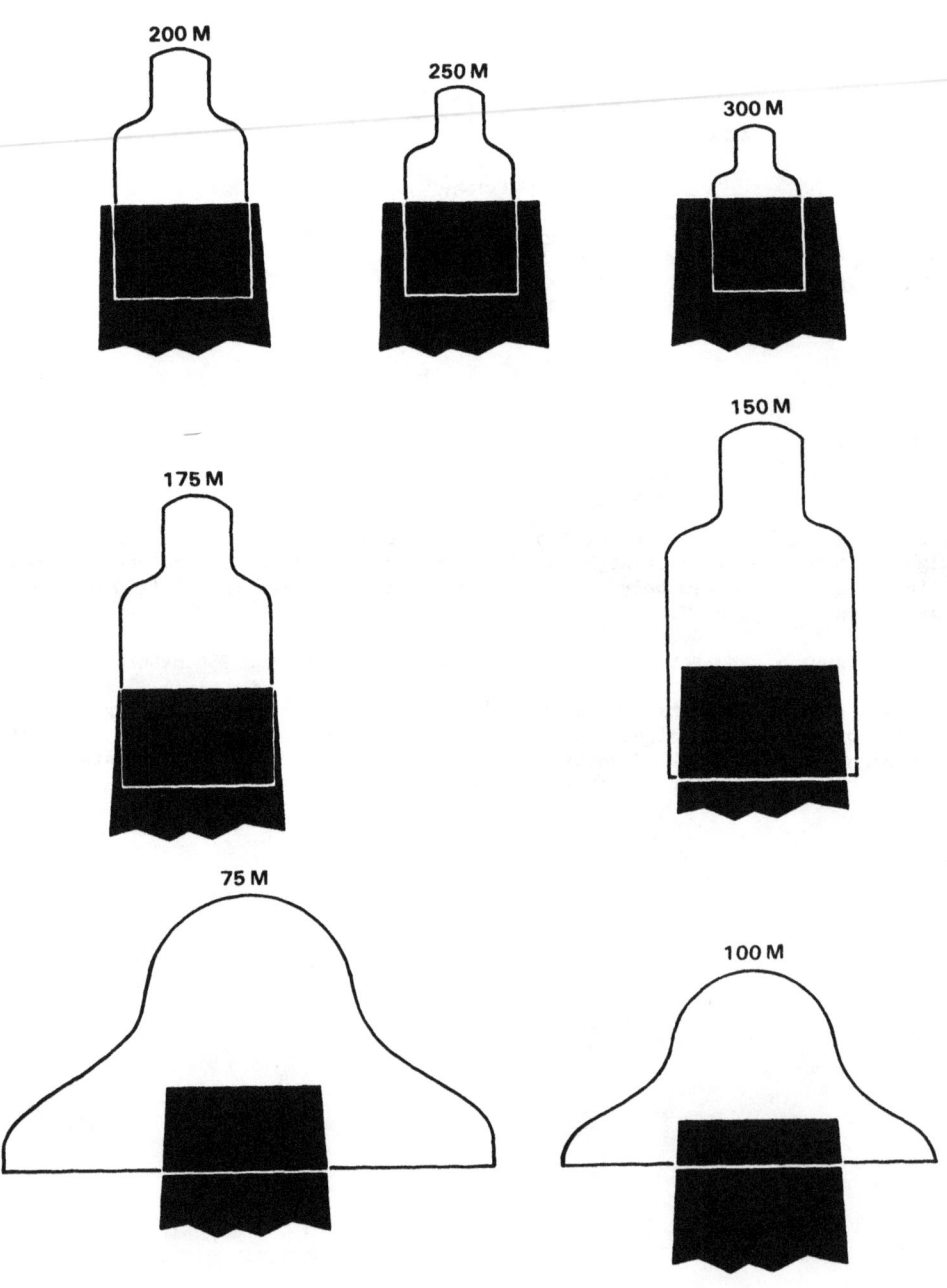

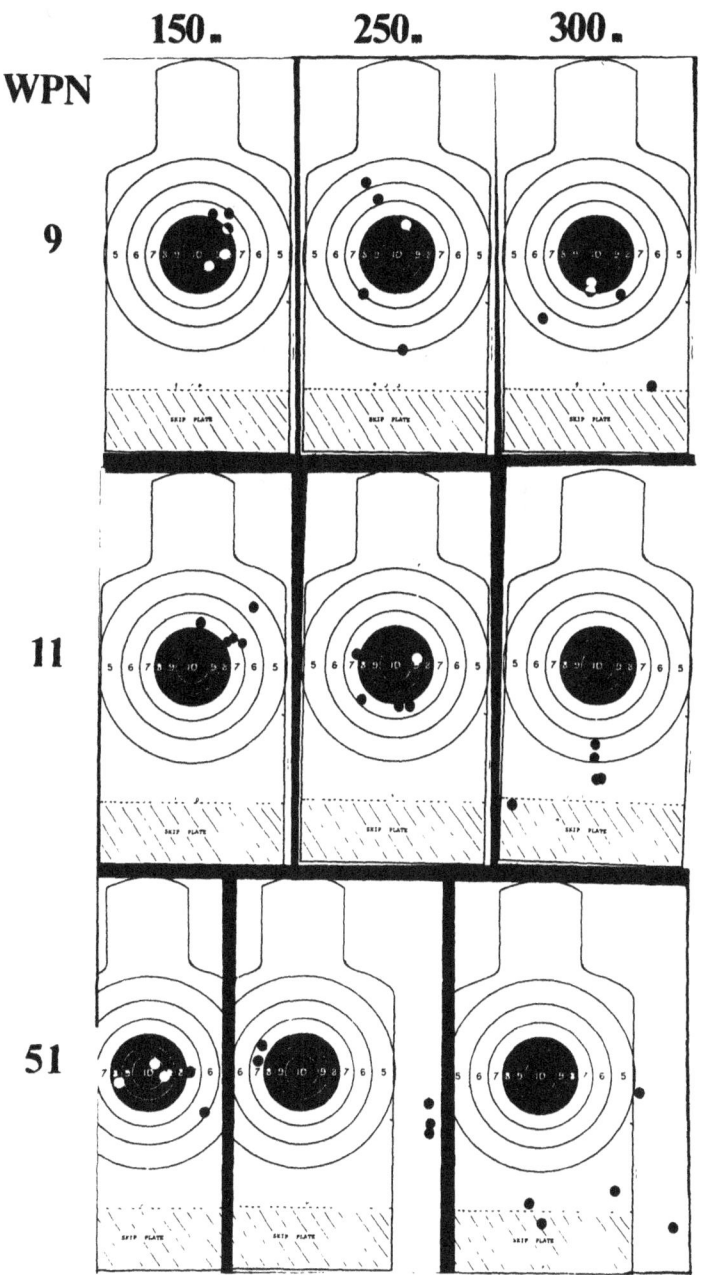

These rifles, zeroed for 250 meters, were aimed at the center of the bull's-eye for each range. Note that the bullets are generally high at 150 meters, centered at 250 meters, and low at 300 meters. The standard E-type silhouette outlines have been added to show shot group size and relevance to field fire performance. Also note that weapon 51 is erratic and must be turned in for repair.

Figure 8. Trajectory--150, 250, and 300 meters (cradle firings).

WIND

For practical purposes, the same gravitational forces are at work all the time. However, the effects of wind will vary greatly based on changes in wind speed and direction. While wind is not a major consideration at close range, it takes on added significance beyond 150 meters. A 10-mph wind is not uncommon. A 10-mph wind blowing directly from the side across the range (a full value wind, see Figure 9) will move the bullet as indicated in Figure 10.

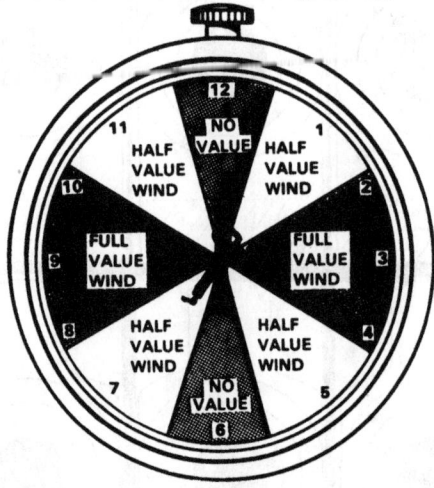

A wind blowing from right to left is called a 3 o'clock wind.

Figure 9. Wind direction by the clock system.

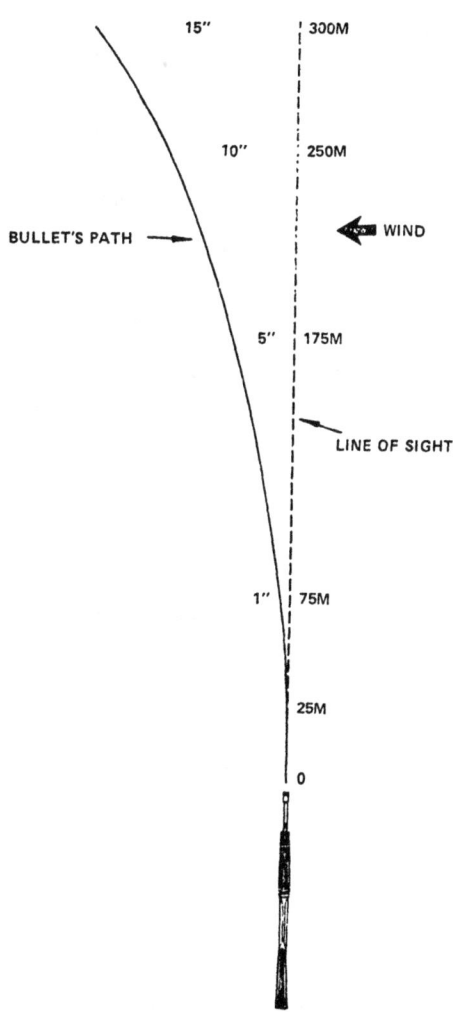

Wind is not a problem when firing at targets at close range. However, it can cause a firer to miss distant targets. (See Figure 5.) A 10-mph wind is not very strong. But, when it is blowing directly from the side, it can bend the path of the M16 round.

As you can see, the wind has little effect on this path inside of 75 meters. However, it has an increasing effect on the bullet's path beyond 75 meters. Thus, while the bullet moves only 1/2 inch between 0 and 50 meters, it drifts 5 inches between 250 and 300 meters.

Figure 10. Effect of a 10-mph wind on bullet's path.

As clearly reflected in Figure 10, wind is not a problem at close ranges, a full value 10-mph wind from 3 o'clock moves the bullet only 1/10 inch at 25 meters and less than 1 inch at 75 meters. However, as distance increases, the wind is a definite consideration. Note that a 10-mph wind will move the bullet 5 inches at 175 meters--requiring hold-off or 2 to 3 clicks of windage adjustment to hit target's center. Allowance for wind must be considered at ranges beyond 150 meters. One should also note that the wind curve, much like the bullet drop curve, does not change uniformly in relation to range--that the effects of wind at longer ranges are much greater than at shorter ranges. However, the wind effect is uniform in relation to wind speed--at 175 meters a 5-mph wind moves the bullet 2 1/2 inches; a 10-mph wind, 5 inches; a 15-mph wind, 7 1/2 inches; a 20-mph wind, 10 inches, etc.

A half value wind will effect the bullet approximately half as much as a full value wind. That is, a 1 o'clock wind having a velocity of 10 mph is equivalent to a 5-mph 3 o'clock wind.

If a wind gauge is not available, there are three common field expedient methods of determining wind velocities.

    1. <u>Flag Method</u>. If a shooter can observe a flag (or any cloth-like material similar to a flag) hanging from a pole, he should estimate the angle (in degrees) formed at the juncture of the flag and the pole. Dividing this angle by the constant number 4 will give the wind velocity in mph (Figure 11).

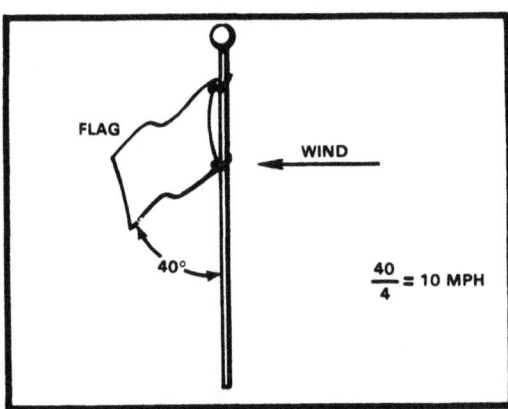

Determine the angle between the pole and the flag and divide by 4 to obtain wind speed in miles per hour.

Figure 11. Determining wind velocity by the flag method.

2. _Pointing Method._ If no flag is visible, a piece of paper or other light material may be dropped from the shoulder. By pointing directly at the spot where it lands, the angle (in degrees) can be estimated. This figure is also divided by the number 4 to determine the approximate wind velocity in miles per hour (figure 12).

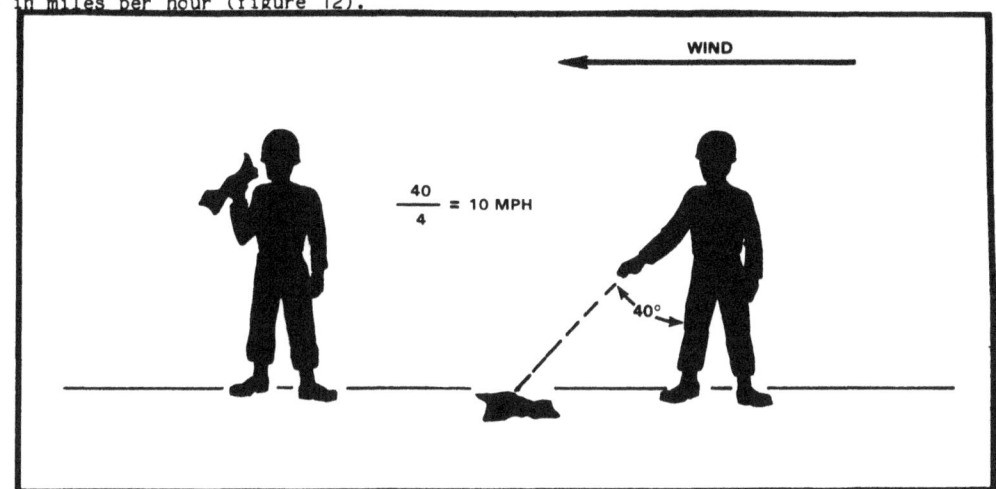

Determine the angle and use the same procedure used in the flag method--divide the angle by 4.

Figure 12. Determining wind velocity by the pointing method.

3. _Observation Method._ If the above methods cannot be used the following information will assist in determining wind velocities:

    a. Under 3 mph, winds can hardly be felt, but the presence of slight wind can be determined by drifting smoke.

    b. A 3- to 5-mph wind can be felt lightly on the face.

    c. Winds of 5 to 8 mph keep tree leaves in constant motion.

    d. At 8 to 12 mph, winds will raise dust and loose paper.

    e. At 12- to 15-mph will cause small trees to sway.

Given the nature of the record fire course and the nature of combat, sights will normally not be adjusted for wind. Therefore, the hold-off technique must be used to compensate for the effects of wind. The placement of the front sights to allow for wind is shown in figure 13. During BRM, the trainer should estimate the impact of wind and tell the soldiers how much to adjust the aiming point.

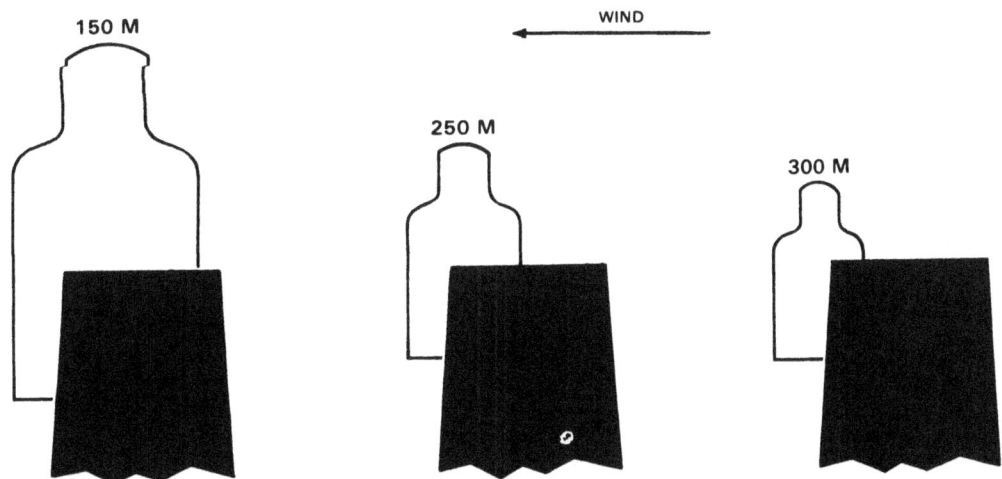

Shown is the placement of the front sight that will bring bullets to target center when firing in a 10-mph full value 3 o'clock wind.

Figure 13. Adjusted aiming point--wind.

THE RIFLE

CONFIDENCE

A very important aspect of any marksmanship program is to develop soldier confidence in the service rifle. The M16 rifle was issued to US combat troops in Vietnam because it was proven to be the best combat weapon for that type of low intensity conflict. Following an evaluation of the M16's performance in combat and additional testing, the decision was made to adopt the M16A1 as the standard service rifle.

There has been considerable controversy over the adoption of the small bore (5.56-mm) rifle in lieu of the large bore (7.62-mm) M14 rifle. However, several other major military powers, to include the Soviets, are going to a 5.56-mm type cartridge. It is possible that this controversy over the M16 versus the M14 has had some negative impact on the marksmanship program, resulting in a lack of confidence in the M16 rifle. The marksmanship program trainer should be aware of the facts and contribute to producing soldiers that have confidence in the M16 rifle. The M16 rifle was adopted by the US Army because it was proven to be the best combat weapon available.

Most advantages of the M14 are not advantages at all when the service rifle is employed in accordance with current doctrine--the engagement of personnel targets out to a range of 300 meters. The major advantage of the M16 are:

1. It is lighter and easier to handle.

2. The ammunition is lighter (allowing almost twice as much ammunition to be carried within the same weight limitations).

3. It makes less noise.

4. It has less recoil.

5. It is easier to train all soldiers to fire.

6. It is more accurate out to ranges of 300 meters.

7. The bullet will yaw or tumble in flesh--producing tremendous killing power.

8. It has adequate knockdown power at all ranges (HAVING MORE KNOCKDOWN POWER AT 300 METERS THAN A .45 PISTOL DOES AT THE MUZZLE).

9. It is capable of killing all personnel targets well beyond 300 meters (having the force to penetrate a steel helmet at ranges out to 500 meters).

10. The "flat" trajectory round allows the teaching of one aiming point to obtain target hits at all ranges out to 300 meters.

11. The pistol grip and light weight allows firing with one hand in close-in fighting or fighting in built-up areas.

12. The charging handle offers versatility (making it easy for right- or left-handed firers to operate the weapon).

13. The construction is simple which permits complete field stripping in a few seconds.

It is impossible to adopt a service rifle that everyone believes is the ideal weapon. The M16A1 is not without its faults; however, it has no serious shortcomings. The M16A1 is a lightweight, reliable weapon, capable of delivering accurate semiautomatic fire or a large volume of automatic fire. When cost, performance, reliability, versatility, and trainability are considered, the M16A1 rifle is the best military rifle in the world today.

ACCURACY

An effective way of destroying an individual's confidence in a weapon is to issue him a weapon that will not shoot. Concerned about the accuracy of the typical weapon used by soldiers caused 60 entry-level soldier weapons to be selected at random and tested[3] for accuracy. Weapon quality has been discussed and the targets in figure 1 emphasize the importance of the unit augmenting the direct support maintenance serviceability checks with firing tests of their own.

The 60 test weapons were fired from a cradle, a vise-like mechanism that holds the weapon during firing, and by two individuals, each firing from a prone unsupported position. The targets in Figure 14 provide an indication that weapons No. 9 and 11 are quality weapons while No. 51 is not an acceptable weapon to be issued to a soldier. The cradle firings of the 60 weapons at a range of 25 meters produced the following shot group sizes:

Group size in Centimeters	Number of Rifles	Percent of Total
2 or smaller	36	60
2.1 to 2.9	15	25
3 to 4	7	12
over 4	2	3
	60	100

The weapons that fire shot groups in excess of 4 centimeters in size should not be issued to soldiers before being repaired. The weapons firing shot groups in size from 3 to 4 centimeters are questionable, and soldiers using these weapons should be observed closely for possible problems associated with the weapons.

---

[3]See bibliography.

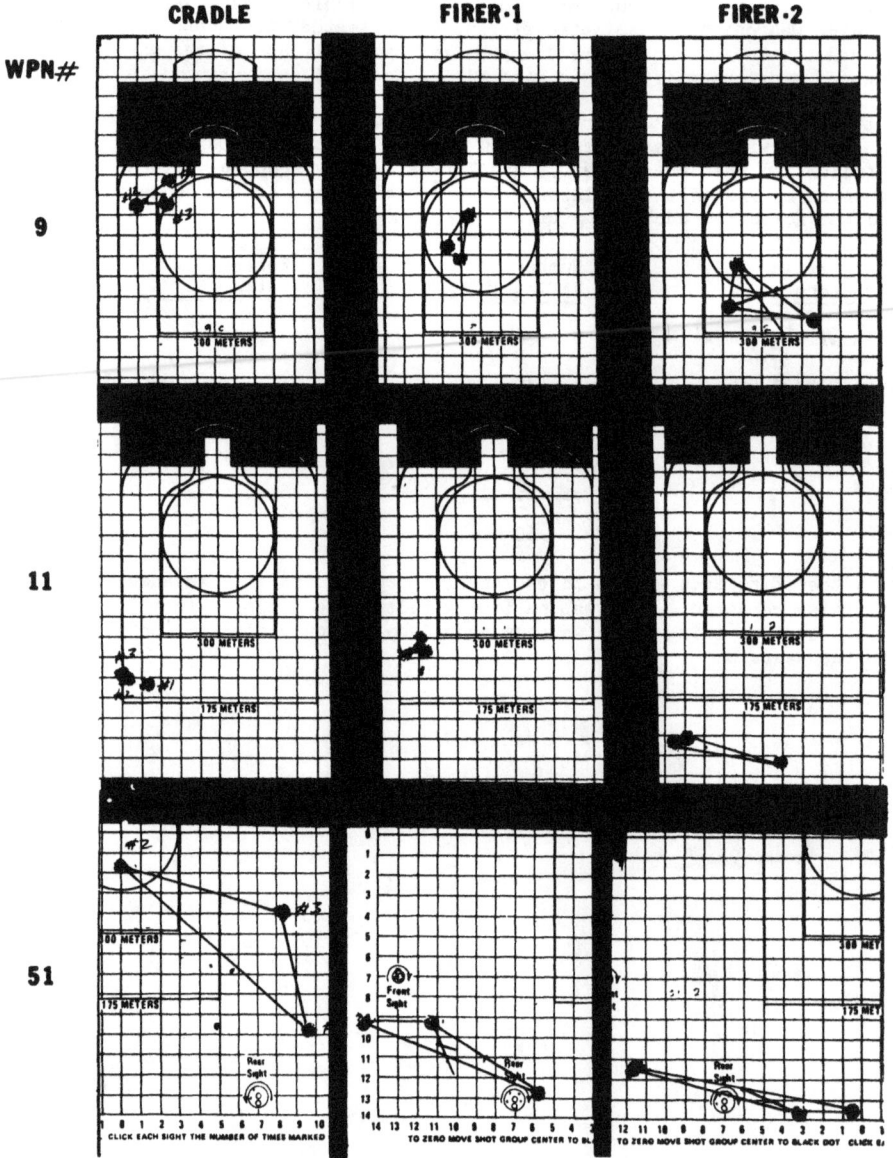

These rifles (3 of 60) were fired from a cradle (weapon vise) and by two firers from prone unsupported positions on a 25-meter range. Rifles passed all serviceability checks before firing. A firing check is the only procedure available to identify bad weapons like No. 51.

Figure 14. Targets used for test firing.

The soldier must have confidence in the weapon. The first step in building this confidence is to issue the soldier a weapon that you know is capable of hitting all targets that will be presented.

CARE AND CLEANING

This Guide makes no attempt to address the details of care and cleaning, since no recent research has been conducted in this area, the subject is adequately covered in current publications. The following suggestions are provided:

1. Emphasize the need to keep the weapon, to include ammunition and magazines, clean by keeping it out of the dirt and sand.

2. Give cleaning priority to the chamber and bolt carrier group. These are the areas that will cause most malfunctions.

3. After zeroing, insure that sights are not moved during cleaning.

4. If the weapon is firing properly, it is recommended that the internal parts not be cleaned after the final confirmation of zero until record fire has been completed. Cleaning will not alter the zero, however, improper lubrication or assembly may cause malfunctions.

The M16 rifle, in good condition and properly maintained, will fire without malfunctions. Any malfunctioning weapon indicates improper maintenance procedures or a weapon that should be inspected by an armorer. A weapon malfunction during record fire should be a rare occurrence.

OPERATION

The operation and functioning of the rifle will not be discussed in detail; however, the marksmanship trainer must be familiar with the operation and functioning of the rifle. This information is available in current publications. The following only serves to highlight some critical points.

The marksmanship trainer should not be concerned with the functioning of the weapon from a technical point of view, but should be concerned with functioning as it relates to teaching the soldier to correct malfunctions.

1. Feeding. To prevent a malfunction in the feeding of the first round, the magazine should be checked to insure that the top round is straight, and the magazine should be locked in the weapon. A hard blow on the bottom of the magazine may dislodge the first round, causing it to misfeed or double feed. Another condition that may cause a malfunction on the first round is putting too many rounds in the magazine. To correct a feeding problem, the weapon and the magazine must be checked. Most feeding malfunctions are caused by bad magazines, requiring cleaning or replacement.

2. Chambering. Pulling the charging handle fully to the rear and releasing it or releasing the bolt catch should fully chamber the round without assistance. The rifle should be visually checked to insure that the first round is seated in the chamber, using the forward assist if necessary to insure that the bolt is locked.

3. Firing. If the hammer falls and the weapon fails to fire, check the cartridge primer. If there is no dent in the primer, the firing pin is probably broken or missing. If the primer is slightly dented, it indicates a weak hammer spring or excessive wearing of the firing pin or bolt area.

4. Extraction. If the cartridge fails to extract, it indicates a dirty chamber, a broken extractor, a weak extractor spring, or insufficient gas supplied to cause proper operation. Too much oil in the chamber can create a vacuum and also prevent extraction.

5. Ejection. Failure to eject normally indicates a broken or worn ejector or weak ejector spring.

IMMEDIATE ACTION

Immediate action is the unhesitating action taken to correct a stoppage without an investigation of the cause of the stoppage. The procedure is to tap the magazine twice (if a round is hung up due to a failure to chamber, this may cause the bolt to ride forward and the soldier can fire if he hears the bolt go forward). If the bolt does not go forward, pull the charging handle to the rear and look into the chamber. If the chamber is clear, release the charging handle, hit the forward assist, and attempt to fire. Soldiers should practice repeatedly to develop the immediate action response so that a simple stoppage will not cause them to fail to engage more than one target on the record fire range.

REMEDIAL ACTION

Remedial action is the action taken to correct a weapon malfunction when immediate action has failed to correct the problem. Initial-entry soldiers should be given assistance with remedial action, and efforts should be taken to prevent a recurrence.

BASIC SHOOTING FUNDAMENTALS

Two things are required to hit targets: the rifle must be aligned with the target, and the hammer must fall without without disturbing the lay of the weapon. The soldier can accomplish this by placing the front sight post on the target and squeezing the trigger.

It is important to understand and focus on the critical elements of the firing process--on those elements that have major influence over the strike of the bullet. If you can teach your soldiers to put the front sight post on the target and squeeze the trigger, they will all hit at least 23 of the 40 record fire targets. This can best be accomplished through teaching the four basic fundamentals:

1. Steady position.

2. Aiming.

3. Breath control.

4. Trigger squeeze.

These basic fundamentals form the heart of the BRM program. It is very important that you understand and believe in these four fundamentals. The current program does not represent drastic changes in marksmanship, but there are some significant changes in emphasis. For example, sight alignment is relatively unimportant, the Shot Group Analysis Card is considered invalid and has been deleted, and the eight steady-hold factors are not taught.

STEADY POSITION

Before the soldier can hit targets, the weapon must be held in a steady position. This is the first shooting skill the soldier should master, and it should be mastered before firing the first live rounds.

Supported Position. Soldiers should be encouraged to always fire from supported positions if at all possible. Any type of support (tree, truck tire, rolled up field jacket, etc) will greatly assist shooting accuracy. All supported firing in the BRM program is performed from the foxhole position with sandbag support. This is a stable and comfortable firing position, making it easy for the novice firer to master shooting skills quickly. It is critical that the weapon be fully supported by the sandbags. The nonfiring hand and arm should support no part of this weight. The firing elbow must be firmly planted on the ground outside of the foxhole position. In effect, the front of the weapon is supported by the sandbags (the nonfiring hand is between the weapon and the sandbags for the purpose of controlling the weapon for subsequent shots) and the rear of the weapon is firmly supported by the firing hand (pistol grip) and the pocket of the firing shoulder (rifle butt). The firing hand and shoulder are steady because they have straightline bone support to the firing elbow, which is firmly planted on the ground outside the fighting position.

Unsupported Position. In some situations, the soldier may be required to engage targets when no support is available. Therefore, the soldier must learn to hold the weapon steady with only bone support. The BRM program uses the prone unsupported position, the steadiest of all unsupported positions. Since half of the record fire targets are engaged from the prone unsupported position, it is critical that soldiers be proficient from this position. A key to this position is getting the nonfiring elbow as far under the weapon as possible. This provides for the greatest amount of bone support and will allow the soldier to maintain a steady position for an extended period.

The supported fighting position and the prone unsupported position will be discussed in detail later.

It is important to build confidence in the soldier's ability to fire the rifle. Therefore, it is important for the soldier to have mastered steady positions before firing the initial rounds. If a soldier has not yet learned to hold the weapon steady, you will be wasting your time and his by going on to other fundamentals. Learn to hold the weapon steady, and the remainder of the shooting process will be much easier to master.

AIMING

Having mastered the task of holding the weapon steady, the weapon must be aligned with the target—**exactly the same way for each firing.** The peep sight on the M16 is a superior sighting system. The peep sight is located well to the rear of the receiver, placing it close to the eye, thereby increasing sight radius (distance between front and rear sight). The major advantage of the peep sight is that it requires the soldier to align only two objects—the front sight and the target.

A good firing position places the eye close to and directly behind the rear peep sight. When the eye is focused on the front sight post, the natural and instinctive ability of the eye to center objects and to seek the point of greatest light (which is the center of the rear sight hole (aperture)) will provide for correct sight alignment. Therefore, the aiming task is simple—<u>place the front sight post on the target.</u>

Among all the other stresses associated with learning a new skill, the soldier has in the past been asked to perform a task that is impossible for the human eye—to focus on three separate planes (objects at three different distances) at once. For the novice shooter, the instinctive and natural ability of the eye will more accurately align the sights than the conscious attempts by the shooter. It is intended that sight alignment be taught, but this subject should be placed in proper perspective and not be presented to the soldier as such an important shooting fundamental.

Relying on this natural ability of the eye to seek the center of the rear sight hole, point of aim becomes more important than sight alignment. To obtain correct point of aim, the tip of the front sight post must be placed on the aiming point. To accomplish this, <u>the eye must be focused on the tip of the front sight post.</u> This may cause the target to appear blurry, while the front sight post is seen clearly. There are two reasons that make it important to focus on the tip of the front sight post. First, if the front sight post is seen clearly, only minimum aiming error can occur while a more significant error can occur if the front sight post is blurry (that is, focusing on the target). A blurry front sight post may cause an alignment error and may cause a target miss. (If the front sight post is damaged, it should be replaced; if it is shiny, it must be blackened.) The second reason for focusing on the tip of the front sight post is that this forces good sight alignment. When the eye is focused on the tip of the front sight post so that it is seen clearly and distinctly, the eye must be centered in the rear sight hole. If the eye moves from the center of the rear sight hole while focusing on the tip of the front sight post, the aperture will begin to darken on one side, and the front sight post cannot be seen as clearly and distinctly as it had been. Therefore, with the focus of the eye directed to the tip of the front sight post, minimum point of aim error will result, and the natural ability of the eye will provide for good sight alignment. It is difficult to recreate what is perceived by the eye looking through the rear sight hole, but Figure 15 attempts to show what the eye should see.

CORRECT

Focus on the tip of the front sight post so that it can be seen clearly and distinctly--and correct sight alignment is automatic.

Figure 15. Sight alignment.

It is suggested that you test this sight alighment theory with a rifle. Assume a good firing position with your eye looking through the rear sight, forget about sight alignment, and focus on the tip of the front sight post. Check the rear sight. You should find that it is aligned. While continuing to focus on the tip of the front sight post, move your eye to the side of the rear sight hole and it will appear to darken and get blurry.

Considerable practice in aiming should be considered before firing the initial live rounds. During dry firing, after the front sight post has been placed on the target, the soldier should change his eye focus back and forth between the front and rear sights to insure that good sight alignment has been obtained. Before firing, the soldier should be confident that placing the tip of the front sight post on the target will provide correct sight alignment, making aiming a simple process. Place the tip of the front sight post on the target.

BREATH CONTROL

It is impossible to maintain a steady position, keeping the tip of the front sight post on the aiming point, while breathing. Therefore, breathing must not be taking place at the moment of firing. the novice firer should be taught to inhale, then exhale normally, and hold his breath at the moment of natural respiratory pause. The shot must then be fired before feeling any unpleasant sensations from holding the breath. As the firer's skills improve, and as timed or multiple targets are presented, the soldier must learn to hold his breath at any part of the breathing cycle. Breath control should be practiced during dry firing exercises until it becomes a natural part of the firing process.

TRIGGER SQUEEZE--THE MOST IMPORTANT FUNDAMENTAL

The three fundamentals discussed above get the weapon into a steady position and correctly pointed at the target. Most novice firers can accomplish this in a relatively short period. But all of this is wasted if the trigger is not properly controlled. The charts previously used to show errors that result from improper sight alignment (not having the front sight post centered in the rear sight hold) do not accurately reflect the sight alignment problem, but they do reflect the seriousness of weapon alignment problems.

Alignment of the overall weapon is critical. The sight radius on the M16 (distance from the front sight to the rear sight) is 50.17 centimeters, or approximately 1/2 meters. Any alignment error between front and rear sights will repeat itself for every 1/2 meter the bullet travels. For example, at the 25-meter line, any error in weapon alignment will be multiplied 50 times. If the weapon is misaligned by 1/10 inch, it will cause a 300-meter target to be missed by 5 feet. The alignment error is normally not a function of improper sighting, but is a function of improper trigger control, causing misalignment of the rifle at the moment of firing. The trigger squeeze is the most important fundamental for the novice firer, because he is most likely to cause weapon misalignment by his poor trigger control.

Trigger squeeze is vitally important in controlling weapon movement. Any sudden movement of the finger on the trigger will disturb the lay of the weapon and cause the shot to miss its intended point. When the trigger is correctly squeezed, the exact moment of weapon firing will be a surpirse to the soldier. The M16 rifle has a low noise level and also low recoil when compared to many other military rifles. But the normal, unconcious reflex action of the novice firer to compensate for the noise and the slight punch in the shoulder is such that he will miss the target if he knows the moment the weapon will fire. A common practice is to tense the shoulder in anticipation of the weapon firing. It is difficult to detect, and many soldiers do not realize they are flinching in anticipation of the weapon firing. The ball and dummy exercise (App A) or the use of Weaponeer are excellent techniques for detecting improper trigger squeeze. When the hammer drops on a dummy round and it does not fire, the soldier's relfexive movement will clearly demonstrate to him, and the trainer, that he is not properly squeezing the trigger. The Weaponeer (App A) is an excellent aid in detecting improper trigger squeeze because of its automatic misfire capability and a feature that allows the aiming point to be observed for the last 3 seconds before firing.

The proper trigger squeeze should start with relatively heavy pressure on the trigger during the initial aiming process. For example, for a trigger that requires 8 pounds of pressure, 4 to 5 pounds should be applied during the aiming process and then additional pressure applied after the front sight post is steady on the target and breathing has been stopped.

From the supported fighting position, there should be no reason to deviate from this procedure. If any significant movement of the weapon can be detected when firing from a supported fighting position, additional work on the position is needed. From the prone unsupported position, a small wobble area will exist. The wobble area is the movement of the sight around the aiming point when the weapon is in the steadiest possible position. If wobble is present, the same trigger squeeze technique should be employed. If the front sight strays from the target any time during the firing process, pressure on the trigger should be held constant and resumed as soon as sighting is corrected. The position must provide for the smallest possible wobble area. Under no circumstances should an attempt be made to quickly pull the trigger while the sight is on the target. The best firing results are obtained when the trigger is squeezed continuously while the smallest possible wobble area is maintained and the actual moment of firing cannot be anticipated.

As skills increase with practice, the time required to squeeze the trigger can be reduced considerably. Novice firers may take 5 seconds to perform an adequate trigger squeeze, but as skills improve, the trigger can be squeezed in a second or less. The important thing to remember is that the firer must not know the instant that the weapon is going to discharge. Otherwise, body reflexes will attempt to compensate and will cause misalignment of the weapon. Once the hammer falls, the bullet (traveling at 3,250 feet per second) will clear the muzzle of the weapon before body reflexes can disturb the lay of the weapon.

The fundamentals the soldier should have in mind when he comes to the firing line are to:

1. Establish a steady position.
2. Put the front sight post on the target.
3. Stop breathing.
4. Squeeze the trigger.

This simple procedure will insure target hits at all ranges.

USING THE FOUR FUNDAMENTALS

Before the presentation of targets, the position must be adjusted to allow for holding the rifle steady. The soldier must raise his head slightly (maintaining cheek contact with the weapon), look over the sights with both eyes open, scan the appropriate area, and insure that the targets are detected the moment they pop up. As the target appears, the head is dropped down, the eye is focused on the target. The aiming point for this particular target is determined as the eye is rapidly placed behind the rear sight hole. Initial pressure is applied to the trigger (approximately half the pressure that it will take to pull the trigger) as the tip of the front sight post is placed on the aiming point and as the eye is focused on the tip of the front sight post. Breathing is controlled and final trigger squeeze begins. Everything is performed rapidly to the point of final trigger squeeze. More and more pressure is applied to the trigger until the weapon fires. The instant of firing cannot be anticipated. When sufficient dry firing is conducted, the soldier will be able to apply these fundamentals in a well-coordinated manner, resulting in excellent shooting performance.

Do not allow yourself to be detracted from these four basic fundamentals. These are the factors that have major influence over the strike of the bullet. Too much emphasis on other points will probably degrade the shooting performance of entry-level soldiers. When possible, test these fundamentals for yourself on a live-fire range. Zero a rifle and engage a 175-meter target. From a steady position, put the tip of the front sight post just below the center of the target, allow for any significant wind, stop breathing, and squeeze the trigger. The target will, of course, be hit. Repeat this procedure but add common errors that you feel may be important, for example, different eye relief, rifle cant, sight misalignment, different ways of holding the weapon, different body positions. If the four fundamentals are practiced for each firing, the target will not be missed. It is also important to remember that when these shooting skills are required in combat, they must be performed under conditions of fear and stress. Insure that the soldiers you train will be effective combat marksmen by properly focusing on the basic fundamentals during training.

FIRING POSITIONS

Two firing positions are used in the basic record fire course. Both are good positions for the teaching of basic marksmanship skills.

Remember that only two actions must be performed to hit a target--the weapon must be pointed at the target and the hammer must fall without the lay of the weapon being disturbed. Practically speaking, it is not necessary to hold the weapon tightly with the nonfiring hand to make it shoot in the semiautomatic mode. If it were possible to suspend the rifle in midair, the same place each time, and pull the trigger without disturbing the aim, the rifle would fire a tight shot group. This is because when the weapon is fired without stress, each bullet is exposed to similar conditions. However, the soldier's job is to fire while maintaining complete control of the weapon--being ready to rapidly engage additional targets. Therefore, the soldier must maintain control of the rifle. The importance of this is to understand that minimum effort is required to make the rifle shoot and that strong-arm tactics will detract from, rather than contribute to, accurate shooting. (The effects of stress on the rifle are shown in Figure 16.)

FOXHOLE POSITION (SUPPORTED)

This is an excellent position that provides a stable firing platform. The depth of the foxhole position and the sandbag support should be adjusted for the height and arm length of each soldier.

Nonfiring Hand and Arm. The handguard should rest in a V formed by the thumb and forefinger of the nonfiring hand, and lie across the heel of the hand. The handguard should not be gripped. The nonfiring hand is used to maintain control of the weapon and assist in the rapid engagement of subsequent targets. The forward weight of the rifle must be fully supported by the sandbags. The nonfiring hand and arm should carry none of this weight.

Rifle Butt. The butt of the stock is placed in the pocket formed by the firing shoulder. A firm placement of the butt into this shoulder will lessen the effects of recoil.

Firing Hand. The firing hand grasps the pistol grip, and the forefinger is placed on the trigger in a manner that will not disturb the lay of the weapon while the trigger is being squeezed. Note that the emphasis on placement of the trigger finger is to be able to pull the trigger without causing movement of the weapon. The tip of the finger keeps the finger clear of the receiver and is good for light triggers. If the trigger is hard to pull, firing performance will improve for many soldiers with the finger inserted at least to the first joint. A slight rearward pressure should be exerted by the firing hand to keep the butt of the rifle firmly in the pocket of the shoulder, minimizing the effects of recoil.

Firing Elbow. The body should be supported by the corner of the foxhole position, placing the firing elbow outside of the foxhole position, blocked against solid support.

Stockweld. The stockweld is the mating of the cheek with the stock of the rifle, which puts the eye close to and directly behind the rear peep sight. To obtain a consistent stockweld, a good guide is to place the tip of the nose within one inch of the charging handle for each firing. Some stockweld should be maintained during target acquisition (some part of the cheek in contact with the stock) to assist in rapidly assuming a firing position.

Relax. A good position will eliminate undue muscle strain or tension which distracts from a steady position.

PRONE SUPPORTED POSITION

The firer assumes a comfortable prone position behind the weapon. The forward part of the weapon is supported by the nonfiring hand and arm. The steadiest positions normally result with the nonfiring elbow as far under the weapon as possible; however, the left and right elbows and the nonfiring hand should be adjustd to provide the steadiest possible firing position. The other factors discussed above apply equally to the prone position.

EFFECTS OF STRESS ON THE WEAPON

Firing tests have revealed that any pressures on the front handguard may have an adverse effect on accuracy unless the same pressure is applied for each firing. The targets in figure 16 show the difference in bullet strike as a result of pressure on the forward part of the weapon. These targets were fired with a hasty sling, which pulls the forward part of the barrel down; and a bipod with downward pressure at the rear of the weapon, which pushes the forward part of the barrel up. This difference in bullet strike will not result from normal hand pressure to the rear, but rearward pressure does not contribute to accuracy. Previous military rifles (M1 and M14) had recoils that discouraged firing without a firm grip on the weapon; however, the light recoil of the M16 and the extra control provided by the pistol grip allows firing without a firm grip on the weapon. Pressure on the handguard will influence bullet strike and is an item that must receive consideration in all firing positions. It is important to zero with the same type of pressure on the weapon that will be used for firing.

The point is that a lack of grip by the nonfiring hand will not detract from shooting performance—too much grip may detract.

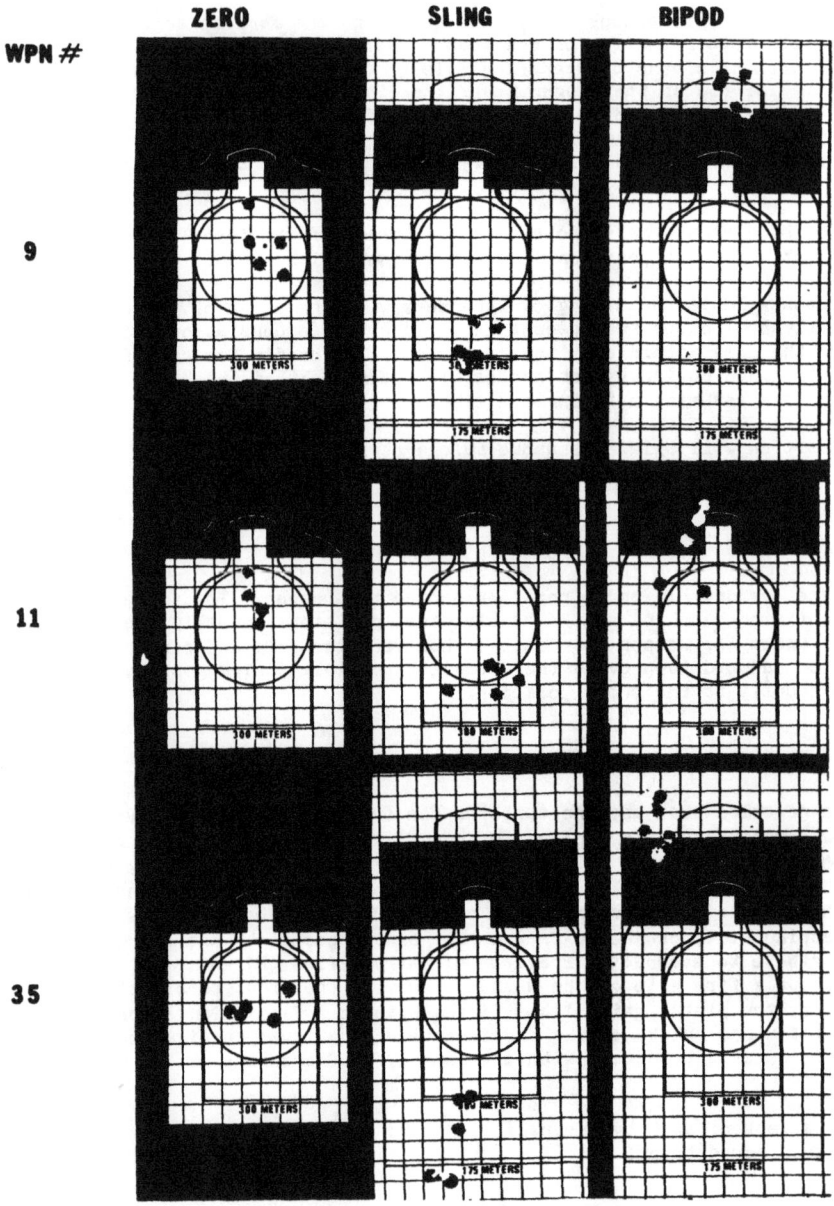

The left targets show the normal zero, the center targets were fired with a hasty sling, and the targets at the right were fired with a bipod attached and downward pressure applied with the left hand forward of the carrying handle. Note that the difference in bullet strike at 300 meters would be 2 to 4 feet.

Figure 16. Rifle stress (handheld).

ORGANIZATION FOR TRAINING

The current program recommends that firing lines have a section designated for each platoon. For example, if there are 36 points being used on the firing line, each of the four platoons would be assigned 9 points. This allows the leaders/cadre of that platoon to concentrate their efforts on the personnel assigned to their platoon. By the end of each period of instruction, the platoon cadre should have a list of their platoon personnel who have demonstrated potential firing problems. An attempt should always be made to diagnose and correct individual problems before the next period of instruction.

REMEDIAL TRAINING

Soldiers who do not obtain the specified minimum standard for any period of marksmanship instruction should have their problems diagnosed and be provided appropriate remedial training. All soldiers must demonstrate a grasp of the basic fundamentals of shooting before allowed to continue in subsequent periods of instruction. A target or scorecard is provided for each firing period of instruction. It is recommended that soldiers retain these as individual progress records and for periodic checks by cadre. These individual performance records will be valuable diagnostic aids.

ZEROING

The battlesight zero is the sight setting which provides the highest probability of hitting the majority of combat targets without sight changes and with minimum changes in point of aim. As shown in figure 17, zeroing the M16 rifle for 250 meters provides excellent hit probability at all ranges out to 300 meters.

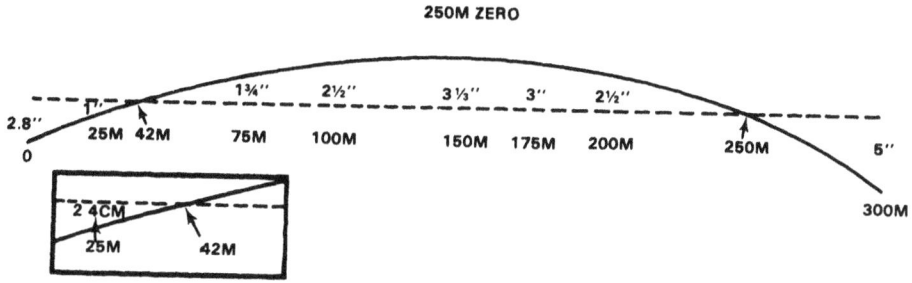

Figure 17. Trajectory--250-meter zero.

The best method of obtaining a 250-meter battlesight zero is to fire at a range of 250 meters, making the necessary sight adjustments to place the center of the shot group on the aiming point. However, such a method requires extensive terrain and training time for firers to move between the firing line and targets to check the location of their shot groups or the use of ranges and/or target equipment not currently available at most installations. When projectiles of the same type and caliber are fired from the same type rifle they have the same general trajectory. The rifle can be zeroed at any known distance along the trajectory course. Referring to figure 17, if the bullet's strike is adjusted to hit 1 inch below point of aim at 25 meters, point of aim at 42 meters, 1 3/4 inches above point of aim at 75 meters, etc., an acceptable 250-meter battlesight zero will result. The previous practice has been to zero on the 25-meter range, adjusting bullet's strike 1 inch (2.4 centimeters) below point of aim.

The procedure of aiming one place and adjusting bullets to hit another has been confusing to many soldiers. Marksmanshp research revealed that using the long-range sight and adjusting bullet strike to coincide with point of aim at 25 meters would produce a good 250-meter zero when the sight was flipped back to the regular aperture. The standard rear sight of the M16 consists of two apertures (holes). When the unmarked aperture (regular sight) has been zeroed for 250 meters, flipping to the aperture marked "L" automatically provides a zero of 375 meters. Figure 18 shows the difference in bullet trajectory between the regular and long-range sights. As discussed above, the rifle may be zeroed at any point along the trajectory curve produced by the regular or long-range sight.

Therefore, by flipping to the long-range sight and adjusting bullet impact to coincide with point of aim at 25 meters produces a 375-meter zero on the long-range sight; or a 250-meter zero when the sight is flipped back to the regular aperture.

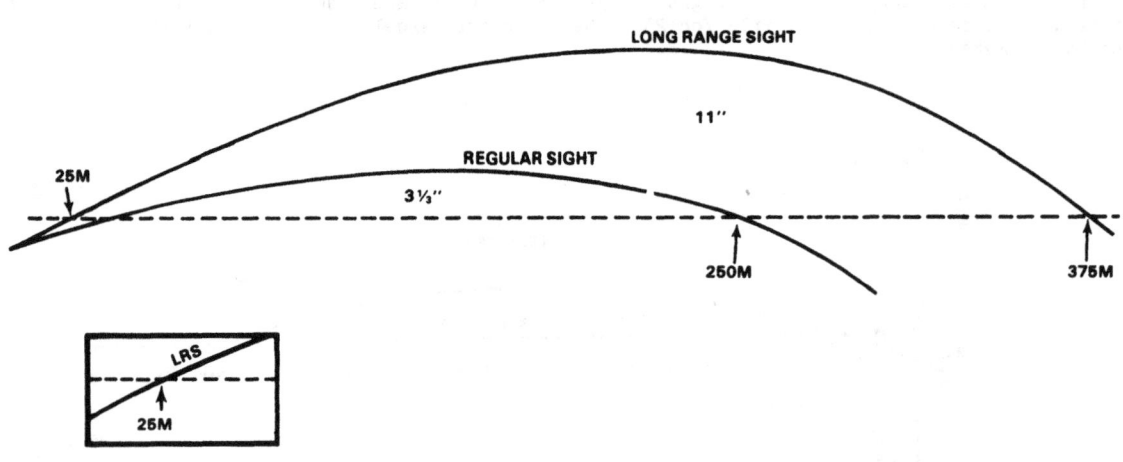

After the standard sight has been zeroed for 250 meters, flipping to the long-range sight results in a zero for 375 meters. Note that the trajectory, when zeroed for 375 meters, crosses the line of sight at 25 meters. Therefore, by flipping to the long-range sight and adjusting bullet impact to point of aim at 25 meters, the standard sight will be zeroed for 250 meters.

Figure 18. Trajector--long-range sight.

The canadian bull presented another prolem, in that when zeroing was completed (the soldier having learned to aim at the bottom of the target), all subsequent firing was at silhouette type targets (with a center-of-mass aiming point). It was found that the quality of zero did not deteriorate, and much confusion was eliminated, when a scaled silhouette target was substituted for the Canadian bull. The new zeroing target in figure 19 allows a soldier to shoot a scaled silhouette target (providing the same visual perception as an actual silhouette target at a range of 250 meters) and adjust bullet strike to coincide with a center-of-mass aiming point. The new zero target also provides additional information to assist in making sight changes. To determine sight changes, the firer locates the shot grouped center, picks out the horizontal and vertical lines nearest that point, and finds from the margins how many clicks to move which sight and in what direction. The target also shows the soldier where his bullets (fired at 25 meters) would have hit on an actual target at 250 meters. Soldiers can be shown that a tight shot group located on the intersection of the two zero lines at the center of the circle is the desired zero. They can see that the nearer to the dot and the tighter the shot group, the more likely they would be to hit a 250-meter enemy target.

The circle on the silhouette is the maximum allowable shot group size for zeroing. While the object is to shoot the smallest possible shot groups, as close to the center of the circle as possible, a rifle that shoots all three shots inside the circle is considered to be zeroed. The circle is equal to the 19-inch width of the pop-up field fire target at 300 meters. This change from 5.2 centimeters to 4 centimeters (at 25 meters) provides a more stringent, but more valid standard for zeroing.

(Drawings not to scale)

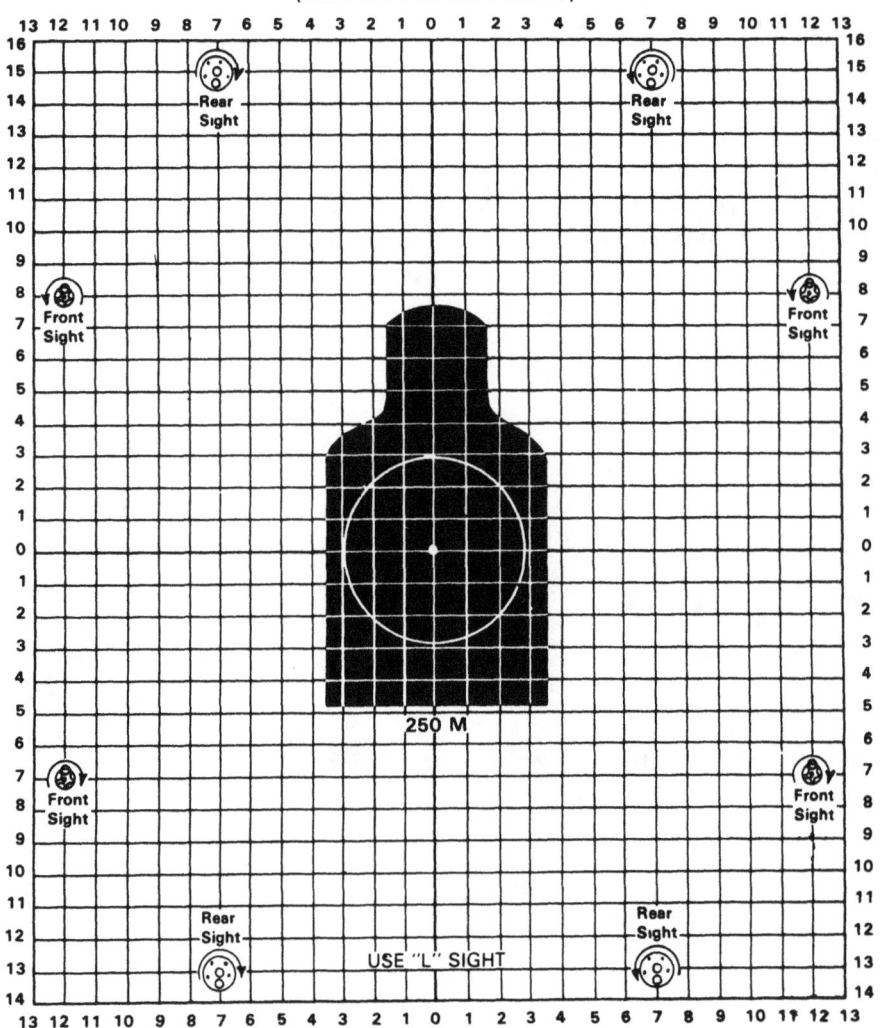

1. Aim at target center, adjust sights to move shot group center as close as possible to white dot.

2. At completion of zero, rotate rear sight to unmarked aperture and weapon will be battlesight zero for 250 meters.

Figure 19. 25-meter zeroing target for M16A1 rifle (with standard sights).

ZEROING PROCEDURE

Previous marksmanship program conducted zeroing before shooting fundamentals were mastered and did not allow for an opportunity to check or confirm the zero during the remainder of the program. Additionally, following the zero period, the soldier was not provided with information concerning target hits and, in most cases, was provided no feedback about target misses. The current program provides for a considerable increase in firing at the 25-meter line, where feedback is provided on the placement of all shots. It also provides a downrange feedback exercise (to be used when know-distances (KD) ranges are not available) where sight changes can be made at longer ranges. There is then a return to the 25-meter line late in the program to confirm and/or refine the soldier's zero.

The primary emphasis for all 25-meter firing should be on the teaching of shooting fundamentals--the shooting of tight shot groups. Once a soldier can shoot well, the zeroing process is simple.

Keeping in mind the variability associated with each round, sight change decisions should always be based on previous shot groups fired as well as on the last shot group. The current program also includes a continuation of firing after the initial zero standard has been met, for the purposes of skill practice and refinement of zero. The objective of zeroing is to get the shot group center as near to the center of the circle as possible. A zero low in the circle is an actual zero for 175 to 200 meters while a zero high in the circle is an actual zero for 275 to 300 meters. A zero anywhere in the circle will provide for target hits at all ranges but a perfect zero will allow for some shooting errors and still enable target hits.

NOTE: Many tactical units have rifles equipped with the low light level sight system (LLLSS). These rifles do not have a long-range sight capability, therefore, special targets have been designed which place bullet strike 2.4 centimeters below point of aim at 25 meters.

SHOT GROUPING

The targets in figure 20 were fired by an experienced marksman from a prone unsupported position. These targets represent the size and shape of shot groups you can expect from soldiers that are applying correct shooting fundamentals from the supported fighting position.

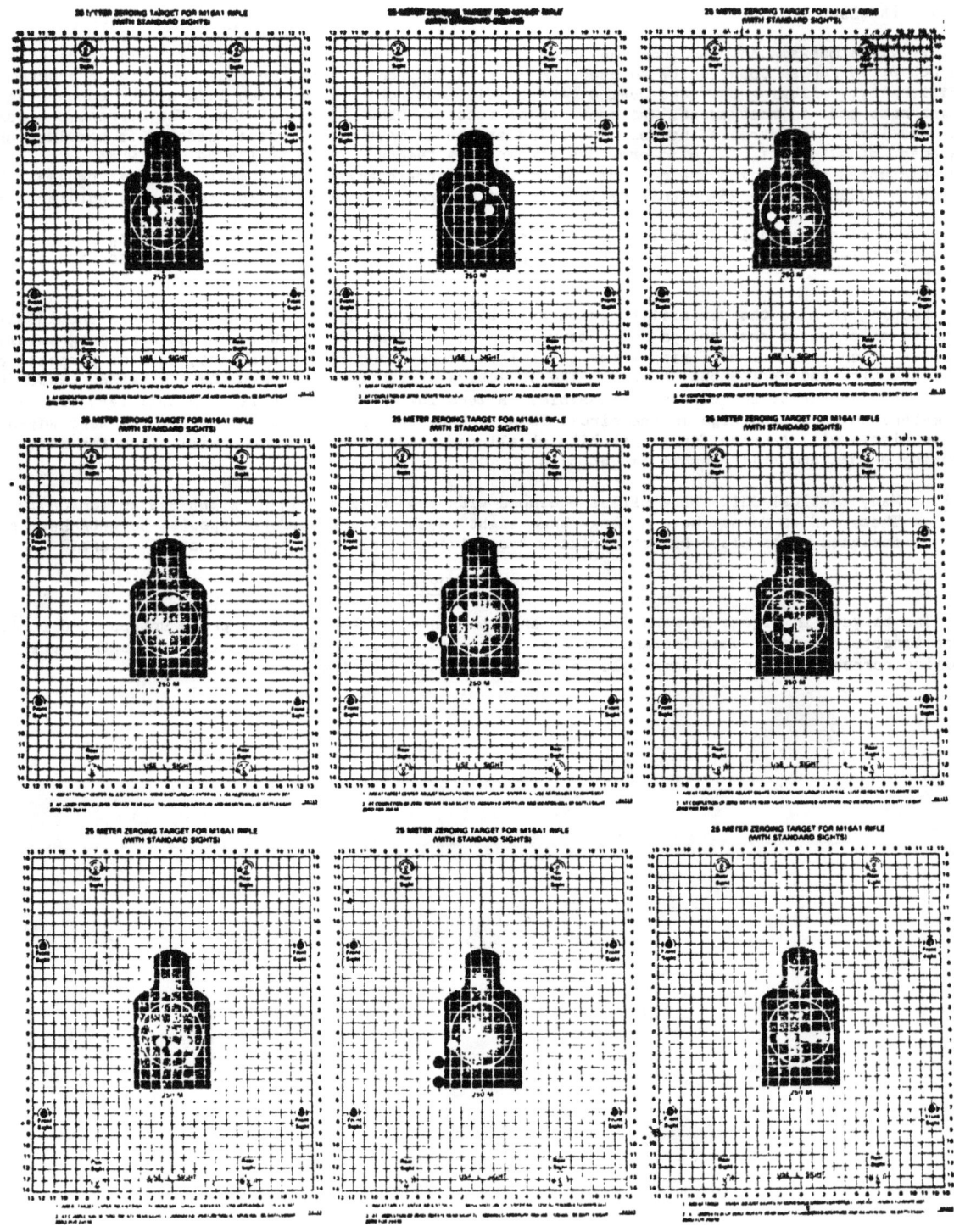

These initial shot groups, fired from nine different rifles, represent what can be expected of entry-level soldiers who correctly apply basic shooting fundamentals. (No attempt was made to zero these rifles for this exercise.)

Figure 20. Shot groups fired from unsupported positions.

The targets in figure 21 were fired by basic trainees with weapons of known shooting quality. The top three targets reflect the correct application of shooting fundamentals.

The bottom three targets obviously reflect firer errors and these are relatively easy to diagnose because 95 percent of these size shot groups are caused by improper trigger squeeze. Only improper trigger squeeze can logically cause this amount of error with a good weapon at 25 meters.

The three targets in the middle of figure 22 are more difficult to diagnose. The soldiers are not properly applying shooting fundamentals. You should observe the soldier firing--check to insure he has a steady and relaxed position, check to insure he understands aiming or check it with the M16 sighting device, insure that breathing is properly controlled, and observe trigger squeeze, or use ball and dummy to insure that the rifle is not being moved at the moment of firing.

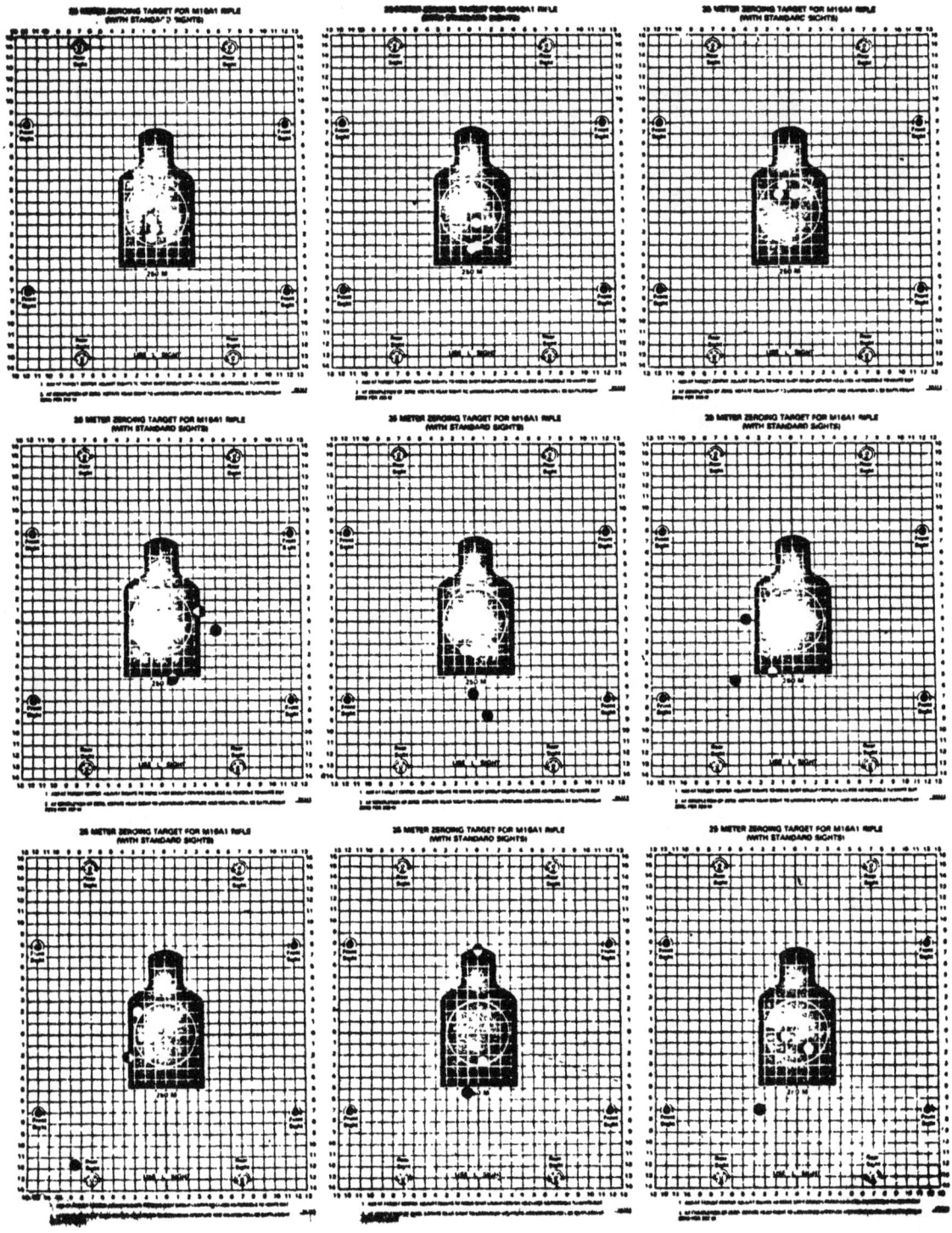

These targets were fired using weapons of known shooting quality. The top targets reflect the correct application of shooting fundamentals. The bottom targets indicate that the soldier knew when the weapon was going to fire and needs to work on tigger squeeze. The center targets indicate that some or all fundamentals are not being correctly applied.

Figure 21. Shot groups from entry-level soldier firings.

The key to correcting firing errors is to stay with the basic fundamentals: a steady rifle with the tip of the front sight post at the aiming point and a surprise firing will result in tight shot groups.

SCALED SILHOUETTE TARGETS

Once the soldier has zeroed his weapon, the trainer is faced with the task of building on and reinforcing basic marksmanship skills. Of key importance is providing a transition to field fire (pop-up) targets, which require new skills on the soldier's part. The trainer must insure that a natural transfer of skills acquired with 25-meter zero targets is provided to the more complex task of engaging targets at range. The use of scaled silhouettes on a 25-meter range is one method of providing this transfer.

The scaled silhouette targets (figure 22--actual target size is 18 x 23) were designed to assist in the transition from firing at 25-meter zeroing targets to engaging pop-up targets on the field fire range. The targets on the left are engaged with three rounds each from the prone unsupported position. This exercise gives an indication, at 25 meters, of likely hits and misses when the firer later shoots at actual pop-up field fire targets. It is also helpful in the teaching and the practicing of proper aiming point.

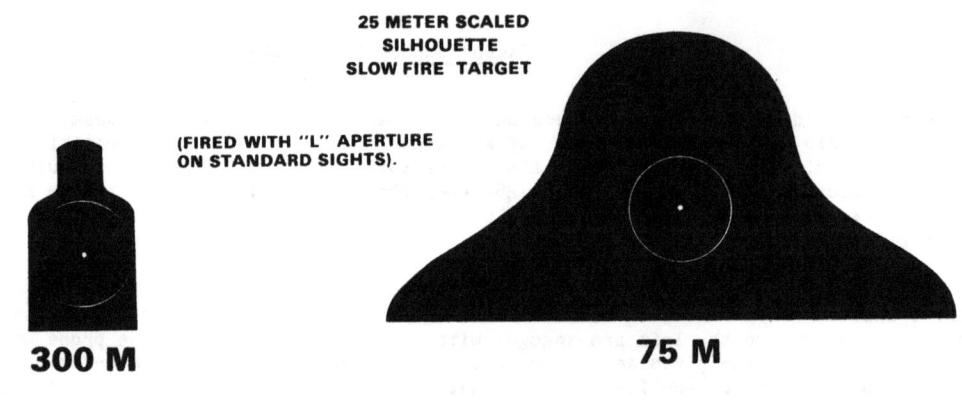

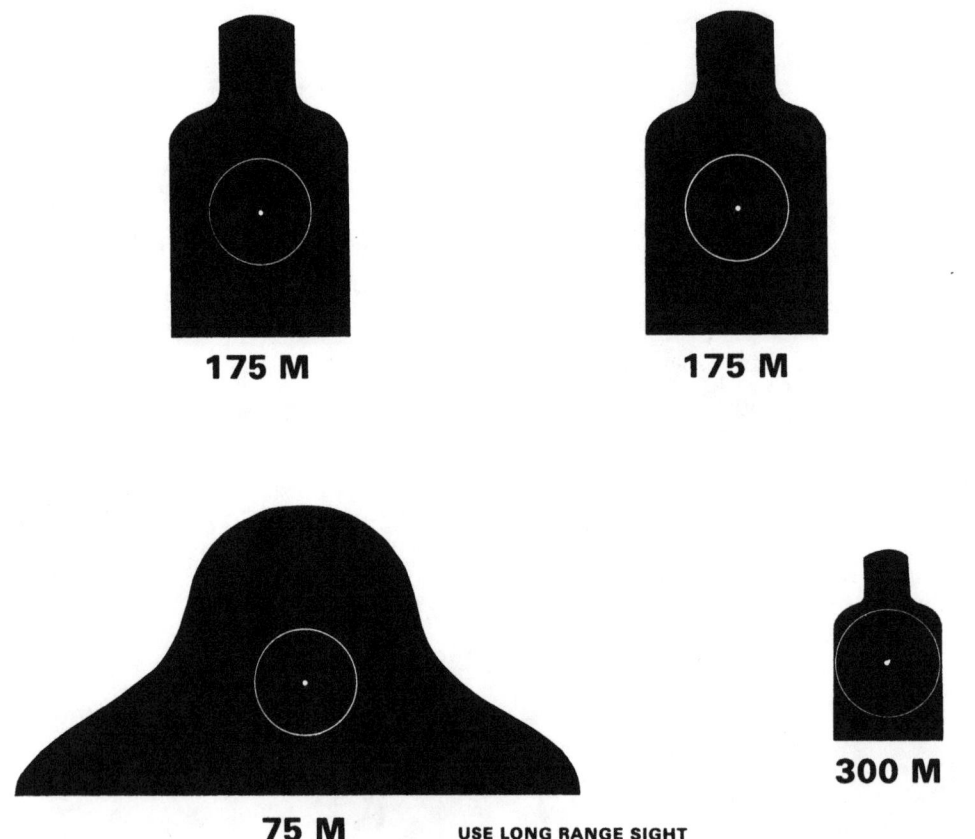

Use long-range sight. The white dot on each target shows the center-of-mass aiming point. Bullets should hit within the circle, but are scored as hits if they hit any part of the silhouette.

Figure 22. 25-meter scaled silhouette targets (slow fire).

Aiming at one of the black silhouettes will result in hitting the silhouette in approximately the same place as an actual size target would have been hit on the field fire range. (Remember that wind and gravity will not affect the bullets much at 25 meters, but will have an effect at greater distances.) These targets have been reduced in size, so that when viewed from 25 meters they appear to be actual size. For example, the firer will note that the front sight appears larger than the 300-meter target, just as it does when a 300-meter target is engaged at actual range. In other words, the visual perception the firer gets from shooting the scaled-down 300-meter target on the 25-meter range is similar to the visual perception he gets from shooting the actual target at a distance of 300 meters.

If these silhouettes are hit, it means that the targets located at the actual distances would probably also be hit. On the other hand, if the silhouette targets are not hit, it is probable that the field fire targets will not be hit. A soldier who cannot hit the 25-meter scaled silhouettes is not ready for field firing.

The timed fire silhouette target at figure 23 uses the same information feedback principle as the 75-175-300-meter scaled silhouette target, but is intended to be a preview of the performance requirements of the combat fire and record fire periods of instruction. The exercise is characterized by:

    1. Inclusion of 50-, 100-, 150-, 200-, 250-, and 300-meter scaled silhouettes to be fired at from the fighting position (supported) and prone position (unsupported).

    2. A requirement to fire at all 10 silhouettes per target under time pressure.

    3. Rapid shifts of point of aim from silhouette to silhouette.

    4. The absence of a clearly defined pattern of silhouettes on the target. Soldiers must quickly choose a strategy for sequencing shots.

    5. An additional opportunity to critique aiming point and teach proper hold for targets at range.

    6. Ability to assess hits and misses for trainer critique.

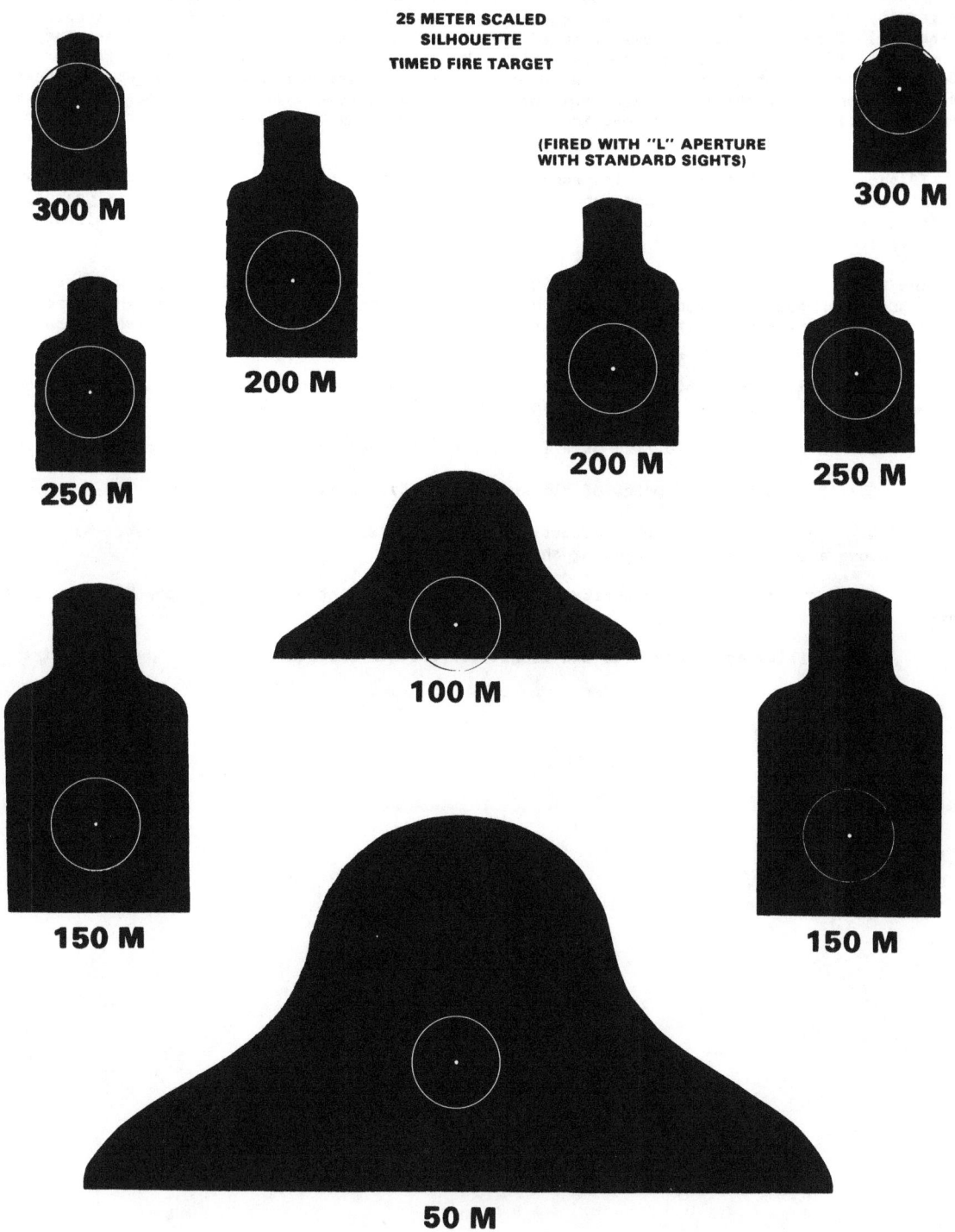

Use long-range sight. The white dot on each target shows the best aiming point for targets at actual distance. If adjusted aiming point is used at 25 meters, bullets should hit within the circles, but are scored as hits if they hit anywhere in the silhouette.

Figure 23. 25-meter scaled silhouette targets (timed fire).

The exercise emphasizes the task requirements of rapid target engagement under time pressure. The soldier completes the exercise with some understanding of what will be required in subsequent exercises. For the more accomplished shooter, the exercise is challenging. For the poorer shooter, the exercise provides another opportunity for diagnosis and remedial training. These training benefits are accomplished on a 25-meter range using an inexpensive, single target format. (Actual target size is 18 x 23 inches.)

An advantage of firing on a 25-meter range is the opportunity for the firer to approach and inspect the target. By locating his shots relative to the point of aim, the firer is provided information on the adequacy of his marksmanship skills. The same principle applies to rounds located on targets at any range. By introducing the soldier to downrange feedback[5] at 75- and 175-meter targets, the trainer provides several additional dimensions to the BRM program of instruction (POI)--

  both hits and misses are captured.

  mistakes are identified and corrected.

  sights are adjusted to provide the best possible zero.

  good shooters are made better.

  spotters are used to help trainers spot soldiers needing help, and effects of gravity and wind are experienced.

---

[5] See bibliography

The use of information feedback as a method of improving skill acquisition has been long substantiated by research studies. This fact has been used to develop a period of instruction of basic rifle marksmanship incorporating feedback of where each round strikes (both target hits and misses). For initial field firing, this is an improvement over the use of only pop-up targets, which offer all-or-no feedback information on hit or miss.

Downrange feedback provides not only the capture of all hits and misses, but also the opportunity to measure how close or far a bullet strike is from where it is intended to be. Provided such information, the soldier can refine his skills. In addition to this logical advantage to downrange feedback, as a practical matter, the exercise is simple to execute. All that is required are large target frames or witness panels on which are stapled E- or F-type silhouettes. Downrange feedback configured in this way does not required extensive investment in equipment.

Targets were developed for use at 75 and 175 meters, including sight change information, and are shown photo-reduced in figures 24 and 25. The actual size of these targets are the same as F-type and E-type silhouettes respectively.

When facilities do not permit this period of instruction to be conducted on a range with targets at 75 and 175 meters, the trainer is still faced with the task of building on and reinforcing basic marksmanship skills. Of key importance is providing a transition of field fire (pop-up) targets which require new skills on the soldier's part. The trainer must insure that a natural transfer of skills (acquired with 25-meter firing) is utilized in the more complex task of engaging targets at range. The use of additional scaled silhouettes on a 25-meter range is one alternate method of providing this transfer.

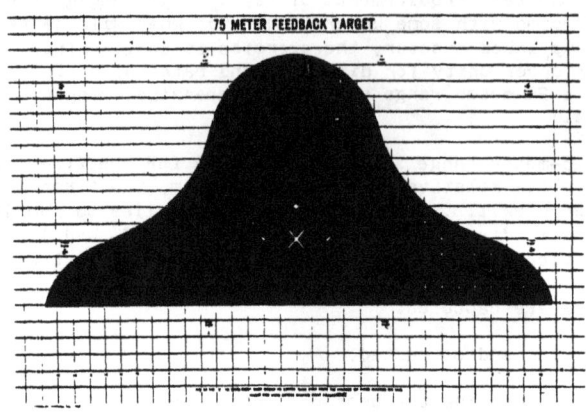

Figure 24. 75-meter known distance target.

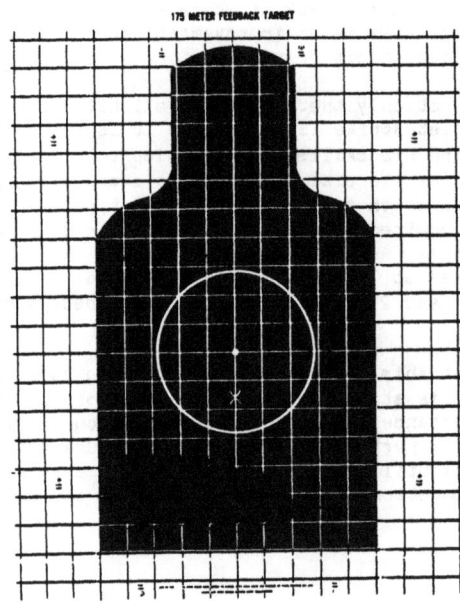

Aim at the X, move shot group to center, click each sight the number of times marked on line. Allow for wind before making sight adjustments.

Figure 25. 175-meter known distance target.

FIELD FIRE

Minor adjustments have been made to the two field fire periods of instruction. These changes are designed to provide a better transition to field fire targets while developing soldier confidence that all targets can be hit, and to present a target array that will better prepare the soldier for record fire.

It is important for the soldier to know where his bullets are going if any targets are missed. A scorer should be used to assist the firer in determining bullet strike for target misses.

TARGET DETECTION

Target detection is included in the current program as concurrent training during the field fire periods. It is listed as concurrent training only to conserve time. Target detection is considered to be an important block of instruction, and it is intended that it be presented as a high priority subject that all soldiers attend. A soldier cannot hit what he cannot detect and thus does not engage.

RECORD FIRE

To qualify on the previous record fire course, a soldier had to hit 17 of 40 targets, or 43 percent. The current program increases the minimum targets hit to qualify from 43 percent to 58 percent, or 23 of 40. This means the soldier must hit six additional targets in order to qualify; however, six targets have been moved to a closer range. Nonetheless, in theory, it is a higher standard because all six of the targets that have been moved to a closer range must be hit. Additionally, many exposure times have been reduced. The basis for this change is threefold:

1. It is universally accepted that the current standard is too low.

2. A recent threat analysis[6] reveals the largest array of personnel targets are normally encountered in the 100- to 200-meter range.

---

[6]See bibliography.

---

3. Firing data from typical M16's indicate that targets out to 200 meters will be hit by all rounds if the soldier applies proper fundamentals. A few targets beyond this range will probably be missed even though the soldier applies all fundamentals that he is taught, but most targets should be hit even at ranges greater than 200 meters. Any soldier who practices the fundamentals will have no difficulty qualifying on this course of fire. Changes in qualification scores are as follows:

	Previous	Current
Expert	28-40	36-40
Sharpshooter	24-27	30-35
Marksman	17-23	23-29
Unqualified	16 and below	22 and below

APPENDIX A

TRAINING AIDS AND PROCEDURES

Several training techniques are available to assist the soldier's understanding shooting fundamentals before live firing or to assist in remedial training for those soldiers who experience problems during training. If newly developed, approved electronic devices (for example, Weaponeer) are available, they are recommended. However, several procedures that should be available at all locations are discussed below--target box exercise, Paige sighting device, dime-washer exercise, M15A1 sighting device, M16 sighting device, dominant eye training aid, and ball and dummy exercise. The ball and dummy exercise is considered to be a good diagnostic exercise that can be used on the firing line to identify the common problem of anticipating weapon firing. The M16 sighting device can also be used on the firing line to assist in diagnosing problems.

The current program includes dry firing as a training procedure. The previous program introduced the soldier to the supported fighting position for the first time during initial live firing. It is considered very important for the soldier to have a period of relaxed time (without the constraints and pressures of the live-fire environment) to practice the basic shooting fundamentals from the foxhole supported and prone unsupported positions before live firing. Dry firing should be conducted until the soldier can effectively demonstrate the performance of basic shooting fundamentals (steady position, aiming, breath control, and trigger squeeze) in a well-coordinated and confident manner. The trainer must insure that dry firing instruction is closely and properly supervised to insure that practice does not become boring. Dry fire training is an effective training procedure which can be implemented throughout marksmanship training, resulting in improved record fire scores.

TARGET BOX EXERCISE

The purpose of the target box exercise is to teach and to permit the trainee to practice consistent placement of the aiming point. The exercise requires a rifle requires a rifle rest, a target box on which is attached a plain piece of paper, a pencil, and a team of firer and target man.

The firer places the rifle firmly in the rifle rest and assumes a modified prone firing position along side of it. While looking through the rifle sights, the firer motions the target man to move the target paddle in line with the sights. It is important to insure that the target box be 25 meters from the muzzle of the weapon, since the silhouette on the target paddle is scaled to depict a 250-meter target viewed at 25 meters. When the silhouette appears in proper position above the front sight post, he gives the signal to mark the paper. The target man marks through the small hole in the paddle. The paddle is then removed and the target man marks the spot #1. The paddle is once again placed at a different starting point against the sheet of paper and the exercise is repeated two more times. After the third "shot", the three points are triangulated and labeled as shot group #1. The exercise should be repeated as many times as necessary to achieve shot groups that will fit into a 4 centimeter circle.

A trainer must monitor the exercise. The trainer should not only observe the trainee as the rifle is sighted, but must discuss shot group size and placement relative to aiming errors. Only when the trainer is satisfied that the trainee is proficient in sighting the weapon should the target man exchange roles and become the firer.

THE RIDDLE SIGHTING DEVICE

The device is used to insure that firers can obtain the correct sight picture with their own weapon. This device consists of two parts:

1. A small metal frame which attaches to the front sight.

2. A plastic strip that has a magnet attached and a silhouette target printed on it.

The device is attached to the front sight post and is then moved until the firer obtains the correct sight picture, which can be checked by a trainer using an M16 sighting device.

THE DIME-WASHER EXERCISE

The dime-washer exercise is an additional activity to perfect the fundamentals of shooting. In addition to the firer's rifle, three other items of equipment are required:

1. A washer or dime.

2. A three-inch plastic disk with a 1-inch center hole.

3. A zeroing target positioned 25 meters from the firer.

The exercise is carried out in two-soldier teams. The firer assumes a supported or unsupported firing position and is critiqued by the assistant and trainer. The assistant assumes a position slightly forward of the muzzle and on the side of the rifle opposite the firer. He places the plastic disk over the barrel and against the front sight guard. The disk serves to block the firer's view of the washer or dime, thus emphasizing correct aiming.

The assistant then positions himself so that he can watch the firer's eye and trigger finger. The firer is now instructed to cock the rifle and aim at the target. When the firer is on target, the assistant balances the washer or dime on the barrel so that it does not touch the flash suppressor. After balancing the washer-dime, the assistant tells the firer, "READY", and closely observes the firer during the trigger squeeze.

If the washer-dime drops from the barrell during the trigger squeeze, the assistant informs the firer of the cause. The rifle is then recocked and the exercise repeated. Other firing positions may be used. The aim of the exercise is for he firer to consistently dry fire without causing the washer-dime to fall.

M15A1 SIGHTING DEVICE

The M15A1 sighting device is a cardboard device designed to demonstrate rifle sight alignment and point of aim. Even though the sight picture of the sighting device is not a completely accurate representation of what the trainee sees when sighting the rifle, the M15A1 is useful during the first sessions of marksmanship fundamentals.

The front of the card has an opening (the rear sight aperture) through which two inserts are visible:

1. The front sight post and blades.

2. A silhouette scaled to the size of the front sight blade. Both are independently movable. Thus, the front sight post can be moved to the center of the aperture, independently of the silhouette which can be positioned above the front sight post. The trainer can demonstrate to the trainee that sight alignment is the process of locating the tip of the front sight post in the central region of the rear sight hole. Although centering the front sight post in the peep sight hole rapidly becomes an automatic process as rifle practice continues, trainees unfamiliar with iron sights may find manipulating the illustratd sight post useful.

The M15A1 also provides for demonstration of the proper relationship between the front sight post and the silhouette. As is the case with the sight alignment, placement of the aiming point should be demonstrated by the trainer who is monitoring the trainee's grasp of aiming fundamentals.

THE M16 SIGHTING DEVICE

The M16 sighting device (also referred to as the Belgian sight or cheater device) is metal with a tinted glass square, that when attached to the rear of the M16 rifle carrying handle, permits a trainer to observe a firer's sight alignment and aiming point.

This device can be used in two ways. The first is to use it while conducting the Paige sighting device exercise. The second is to use the device while dry firing or live firing. However, a brass deflector must also be used during live firing; otherwise, brass cartridge cases ejecting from the rifle will hit the trainer in the face or eyes.

THE BALL AND DUMMY EXERCISE

The ball and dummy exercise may be conducted on any live-fire range. The soldier is provided a magazine loaded with a random selection of live and dummy rounds. In some cases, the exercise may be more effective is the trainee is not informed that the magazine contains dummy rounds so that the normal firing procedure may be observed. Any movement of the body or rifle that can be detected, when the firing pin hits a dummy round, is a clear message that proper trigger squeeze is not being applied.

DOMINANT EYE TRAINING AID

There are several procedures that can be used to determine which eye is the dominant or master eye. The following technique can be performed quickly and is the most accurate field technique available.

Cut a 1-inch circular hole in the center of an 8- x 10-inch piece of material (can be anything from paper to plywood).

The trainer positions himself approximately 5 feet in front of the soldier. The trainer closes his nondominant eye and nolds his finger up in front of and just below his dominant eye to provide the soldier with an aiming point.

The soldier holds the training aid with both hands at waist level and looks with both eyes open at the trainer's open eye. With both eyes focused on the trainer's open eye and arms fully extended, the soldier brings the training aid up between himself and the trainer while continuing to look at the trainer's eye through the hole in the training aid. The soldier's eye the trainers see through the hole in the training aid is the soldier's dominant eye.

WEAPONEER

The Weaponeer is a rifle fire simulator. It operates on 115 volts AC and requires an indoor area approximately 6 feet wide and 18 feet long. The firing station consists of a modified M16A1 rifle tht recoils and provides sound simulation through earphones. The recoil and sound levels may be varied. A misfire mode is available to enable detection of flinching and to reinforce immediate action.

The weapon functions similar to an actual M16A1 rifle and requires similar actions of the firer. The Weaponeer can be erected so that all basic firing positions can be used. The target display consists of a scaled 25-meter zeroing target and E-type silhouetes scaled to represent targets viewed from 100 and 250 meters.

A console contains the controls needed to operate the system. Targets may be manually raised or lowered or set for random raise (variable exposure time) in a killable mode. The console contains a visual display (small TV screen) that shows the target, aiming point of the firer (an excellent way to check for a steady position), and simulated bullet strikes (up to 32). An important feature of the Weaponeer is the playback which shows the aiming point for the last 3 seconds before firing each shot. This is an excellent way to diagnose improper trigger control.

APPENDIX B

BRM POI RECAPITULATION

PERIOD	TITLE	HOURS	ROUNDS
1	Introduction to Rifle Marksmanship and Mechanical Training	4	0
2	Fundamentals of Shooting: Dry Fire	4	0
3	Fundamentals of Shooting: Live Fire	4	9
4	Practice Firing: Zero	8	18
5	Practice Firing: 25-Meter Silhouette	2	18
6A	Downrange Feedback: 75 and 175 Meters	8	30
	or		
6B	Downrange Feedback: 25 Meters	(4)	(36)
7	Field Fire: Single Targets; and Target Detection	4	42
8	Field Fire: Single and Multiple Targets	4	36
9	Zero and timed Fire (25-Meter Silhouette)	4	32
10	Practice Record Fire	4	40
11	Combat Firing/Record Firing	7	90
12	Automatic Firing	2	21
13	Protective Mask Firing	2	20
14	Night Firing	<u>3</u>	<u>30</u>
	TOTAL	60	386
			(See note 1)
	<u>ONLY</u> if 6A is impossible	(56)	(392)

NOTES:

1. Total rounds do not include ammunition required for remedial training, ammunition used for demonstrations, or blank ammunition.

2. When available, the Weaponeer should be used during all periods for diagnostic and remedial training. Soldiers may be returned to remedial training during any period.

APPENDIX C

BIBLIOGRAPHY

1. A. D. Osborne, J. C. Morey, and S. Smith, <u>Adequacy of M-16A1 Rifle Performance and Its Implications for Marksmanship Training</u>, Litton-Mellonics/ARI, Draft Research Report, July 1979.

2. T. J. Thompson, S. Smith, J. C. Morey, and A. D. Osborne, <u>Effectiveness of Improved Basic Rifle Marksmanship Training Programs</u>, ARI/Litton-Mellonics Draft Research Report, January 1980.

3. Osborne, Morey, and Smith, op. cit.

4. Smith, Thompson, Evans, Osborne, Maxey, and Morey, <u>Effects of Down-Range Feedback and the ARI Zeroing Target in Rifle Marksmanship Training</u>, ARI/Litton-Mellonics Draft Research Report, February 1979.

5. Smith, et, al., op. cit.

6. R. D. Klein and T. J. Tierney, <u>Analysis of Factors Affecting the Development of Threat Oriented Small Arms Training Facilities</u>, Litton-Mellonics/ARI Draft Research Report, August 1977.

7. Field Manual 23-9, <u>M16A1 Rifle and Rifle Marksmanship</u>, June 1974.

8. Osborne, Morey, and Smith, op. cit.

9. Thompson, et. al., op. cit.

www.ingramcontent.com/pod-product-compliance
Lightning Source LLC
Chambersburg PA
CBHW081328190426
43193CB00043B/2848